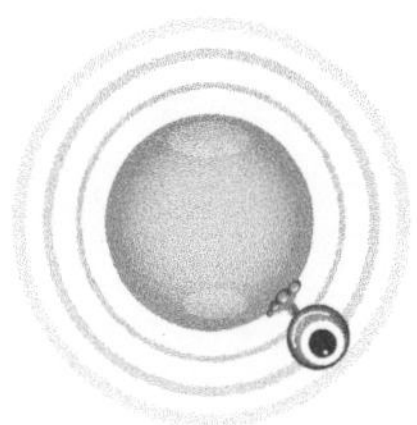

Answer Book

Number Processes
Addition and Subtraction
Multiplication and Division
Fractions

1

Contents

Number Processes Pupil Book 1

Page 3
Hundreds, tens and units

1. 40 + 7 = 47
2. 67
3. 91
4. 173
5. 363
6. 860
7. 902
8. 414
9. 212
10. 6 tens
11. 1 ten
12. 2 tens
13. 9 tens
14. No tens
15. 5 tens

Rocket 11, 22, 33, 44, 55, 66, 77, 88, 99

Page 4
Hundreds, tens and units

1. 361
2. 405
3. 682
4. 240
5. 110
6. 345

In size order: 110, 240, 345, 361, 405, 682

Rocket 36

7. £347 + £100 = £447
8. £281 + £100 = £381
9. £607 + £100 = £707
10. £84 + £100 = £184
11. £840 + £100 = £940
12. £500 + £100 = £600
13. £916 + £100 = £1016

Page 5
Hundreds, tens and units

1. €122
2. €516
3. €300
4. €125
5. €101

Rocket Answers will vary. An example might be 306 +194 = 500.

6. 634, 637, 642
6. Three numbers from 633 to 646.
7. Three numbers from 108 to 116
8. Three numbers from 289 to 300
9. Three numbers from 451 to 454
10. Three numbers from 345 to 371
11. 570, 571, 572

Page 6
10 more, 10 less, 100 more, 100 less

1. a) 430
   b) 550
   c) 580
   d) 620
   e) 660
   f) 730
   g) 790
   h) 850
   i) 860
   j) 700
   k) 990
   l) 970
   m) 820
   n) 690
   1000, 1010, 1020, 1030, 1040, 1050, 1060, 1070, 1080, 1090
2. £154
3. £639
4. £712
5. £439
6. £315
7. £840

The price after £100 has been added:

2. £264
3. £749
4. £822
5. £549
6. £425
7. £950

Page 7
4-digit numbers

1. 6470
2. 3191
3. 4862
4. 9910
5. 1200
6. 8770
7. 4600
8. 5791
9. two thousand and one
10. three thousand three hundred
11. four thousand and fourteen
12. three thousand and three
13. six thousand one hundred
14. one thousand and ninety-nine

Rocket Answers will vary.

Page 8
4-digit and 5-digit numbers

1. 1749, 1769, 1659, 1859
2. 1287, 1307, 1197, 1397
3. 1859, 1879, 1769, 1969
4. 1915, 1935, 1825, 2025
5. 1120, 1140, 1030, 1230
6. 1980, 2000, 1890, 2090

Rocket Answers will vary.

7. 55 642
8. 72 464
9. 84 933
10. 36 728
11. 47 519
12. 66 397

Page 9
5-digit numbers

1. 45 thousands
   3 hundreds
   4 tens
   6 units
   forty-five thousand, three hundred and forty-six
2. 33 thousands
   6 hundreds
   0 tens
   4 units
   thirty-three thousand, six hundred and four
3. 15 thousands
   7 hundreds
   1 ten
   7 units
   fifteen thousand, seven hundred and seventeen
4. 25 thousands
   0 hundreds
   2 tens
   5 units
   twenty-five thousand and twenty-five
5. 60 thousands
   1 hundred
   0 tens
   5 units
   sixty thousand, one hundred and five
6. 32 thousands
   1 hundred
   6 tens
   6 units
   thirty-two thousand, one hundred and sixty-six
7. 14 thousands
   7 hundreds
   0 tens
   9 units
   fourteen thousand, seven hundred and nine
8. 71 thousands
   8 hundreds
   8 tens
   0 units
   seventy-one thousand, eight hundred and eighty
9. 99 thousands
   0 hundreds
   0 tens
   1 unit
   ninety-nine thousand and one

10. 82 thousands
    1 hundred
    9 tens
    0 units
    eighty-two thousand, one hundred and ninety
11. 10 thousands
    0 hundreds
    1 ten
    0 units
    ten thousand and ten
12. 60 004
13. 17 602
14. 80 420
15. 96 047
16. 19 919
Rocket Answers will vary. Possible answers may include: 13 013, 14 014, 15 015, 16 016, 17 017, 18 018, 19 019

Page 10
4-digit and 5-digit numbers

1. 97 899, 97 891, 97 839, 97 831, 97 399, 97 391, 97 339, 97 331, 91 899, 91 891, 91 839, 91 831, 91 399, 91 391, 91 339, 91 331, 57 899, 57 891, 57 839, 57 831, 57 399, 57 391, 57 339, 57 331, 51 899, 51 891, 51 839, 51 831, 51 399, 51 391, 51 339, 51 331
Rocket There are 142 other numbers that children can make if they don't use five cards for each number.
2. 4268, 2846
3. 1234, 3396, 4268
4. 1234
5. 1234, 3412
6. 1234, 1553, 2846, 3412
7. 1553

Page 11
3-digit numbers

1. 561
2. 199
3. 253
4. 472
5. 606
6. 937
7. three hundred and sixty-five
8. four hundred and two
9. two hundred and seventy-nine
10. five hundred and forty-three
11. six hundred and twelve
12. one hundred and eighty-seven
Rocket 963, 852, 741

Page 12
4-digit numbers

1. 1647
2. 3261
3. 6470
4. 2089
5. 5005
6. 3601
7. 7707
8. 4300
9. six thousand, four hundred and seventeen
10. two thousand, three hundred and ninety-nine
11. four thousand and one
12. five thousand, six hundred and seventy
13. one thousand, nine hundred and ninety-nine
14. one thousand and eleven
Rocket one hundred and eleven
two hundred and twenty-two
three hundred and thirty-three
four hundred and forty-four
five hundred and fifty-five
six hundred and sixty-six
seven hundred and seventy-seven
eight hundred and eighty-eight
nine hundred and ninety-nine
one thousand, one hundred and eleven
two thousand, two hundred and twenty-two
three thousand, three hundred and thirty-three
four thousand, four hundred and forty-four
five thousand, five hundred and fifty-five
six thousand, six hundred and sixty-six
seven thousand, seven hundred and seventy-seven
eight thousand, eight hundred and eighty-eight
nine thousand, nine hundred and ninety-nine

Page 13
Comparing numbers

1. 435, 443
2. 82, 88
3. 726, 762
4. 608, 612
5. 541, 548
6. 312, 321
7. 470
7. Any number from 463 to 483
8. Any number from 817 to 860
9. Any number from 346 to 353
10. 320
11. Any number from 211 to 221
12. Any number from 506 to 549
13. 427
14. Any number from 328 to 371
15. Any number from 617 to 660
Rocket 19

Page 14
Comparing numbers

1. 610, 635
2. 350, 380
3. 420, 450
4. 725, 740
5. 555, 565
6. 215, 260
7. Any amount between £491 and £499.
8. About 547 days.
9. Any two numbers between 366 and 371.
Rocket Answers will vary.

Page 15
Ordering numbers

1. 365, 370, 381
2. 402, 408, 412
3. 502, 524, 550
4. 891, 914, 920
5. 734, 745, 754
6. 696, 966, 969
7. 565 cm, 499 cm
7. 565 cm
   Any length from 466 cm to 564 cm
8. 724 cm
   Any length from 625 cm to 723 cm
9. 802 cm
   Any length from 703 cm to 801 cm
10. 399 cm
    Any length from 300 cm to 398 cm
11. 675 cm
    Any length from 576 cm to 674 cm
12. 1098 cm
    Any length from 999 cm to 1097 cm
13. 750 cm
    Any length from 651 cm to 749 cm
14. 532 cm
    Any length from 433 cm to 531 cm
Rocket Any pair from:
910 and 820
820 and 730
730 and 640
640 and 550
550 and 460
460 and 370
370 and 280
280 and 190

Page 16
Ordering numbers

1. 456 g, 460 g
   457 g, 458 g or 459 g
2. 380 g, 410 g
   Any weight from 381 g to 409 g
3. 588 g, 590 g
   589 g
4. 989 g, 999 g
   Any weight from 990 g to 998 g
5. 199 g, 202 g
   200 g or 201 g

6. 646 g, 664 g
   Any weight from 647 g to 663 g
7. 874, 873, 847, 843, 837, 834, 784,
   783, 748, 743, 738, 734, 487, 483,
   478, 473, 438, 437, 387, 384, 378,
   374, 348, 347
8. 157   562
   Any two distances from 158 to 561
9. 635   712
   Any two distances from 636 to 711
10. 241   421
   Any two distances from 242 to 420
11. 78   362
   Any two distances from 79 to 361

Page 17
Ordering

1. Numbers in order: 4160, 4350,
   4540, 4720, 4980
   4160: a
   4350: b
   4540: c
   4720: d
   4980: e
2. Numbers in order: 7721, 7735,
   7746, 7767, 7783
   7721: v
   7735: w
   7746: x
   7767: y
   7783: z
3. Numbers in order: 56 308, 56 333,
   56 345, 56 369, 56 381
   56 308: p
   56 333: q
   56 345: r
   56 369: s
   56 381: t
Rocket 12 numbers between 4000 and
   8000 can be made. 24 numbers
   can be made between 55 000
   and 65 000.

Page 18
Ordering

1. a) £2716   b) £6093
2. a) £8046   b) £8912
3. a) £2593   b) £2675
4. a) £4703   b) £4712
5. a) £33 482   b) £35 276
6. a) £70 001   b) £71 000
7. Answers will vary
Rocket
8. 7632 or 7623
9. 1482, 1428, 1284, 1248
10. Answers will vary.
11. Answers will vary.
12. 45 386 or 45 368
13. 17 632 or 17 623
Numbers as close to 6000 as possible
8. 6237
9. 4821

10. 6234
11. 5987
12. 5864
13. 6123

Page 19
Ordering

1. Portsmath: 29 576, 30 482,
   32 567, 33 291,
   33 910
   Real Mathid: 79 280, 83 460,
   83 490, 83 540,
   86 540
   Math United: 71 936, 72 049,
   72 564, 74 036,
   74 285
Rocket Portsmath:
   32 567 + 33 = 32 600
   32 567 + 433 = 33 000
   Real Mathid:
   86 540 + 60 = 86 600
   86 540 + 460 = 87 000
   Math United:
   72 564 + 36 = 72 600
   72 564 + 436 = 73 000
2. 4740
3. 5835
4. 4800
5. 3550
6. 4300
7. 4748

Page 20
How many hits?

1. 11 400, 11 800, 13 700, 16 400,
   26 000, 45 800, 49 200, 50 000
2. 531, 942, 2780, 4410, 4710, 4820,
   5610, 12 600

Page 21
Greater than, less than

1. $7 m < 11 m$
Answers will vary but should show:
$1 m < 7 m < 8 m < 11 m < 14 m < 16 m$
and $16 m > 14 m > 11 m > 10 m > 8 m >$
$7 m > 1 m$
Rocket Answers will vary.
2. 67 miles
3. 28 miles
4. 77 miles
5. 50 miles
Statements will vary but should show:
28 miles < 50 miles < 67 miles < 77 miles
and 77 miles > 67 miles > 50 miles >
28 miles

Page 22
Greater than, less than

1. $431 < 763$
2. $257 < 384$

3. $902 > 168$
4. $599 < 740$
5. $531 > 325$
6. $650 < 999$
7–14. Answers will vary.
Rocket Answers will vary.

Page 23
Greater than, less than

1. $473 < 521$
2. $357 < 385$
3. $5284 < 5448$
4. $9376 > 9295$
5. $7052 > 6983$
6. $4006 > 3997$
7. $5367 > 5317$
8. $4603 < 4652$
9. $6832 > 6793$
10. $9413 > 9409$
Rocket   1. Any number from 474 to 520
   2. Any number from 358 to 384
   3. Any number from 5285 to
      5447
   4. Any number from 9296 to
      9375
   5. Any number from 6984 to
      7051
   6. Any number from 3998 to
      4005
   7. Any number from 5318 to
      5366
   8. Any number from 4604 to
      4651
   9. Any number from 6794 to
      6831
   10. Any number from 9410 to
       9412
11. 713, 714, 715
12. 897, 898, 899, 900, 901, 902
13. 530, 531, 532, 533
14. 998, 999, 1000, 1001, 1002, 1003,
    1004
15. 3829, 3830, 3831, 3832, 3833
16. 3298, 3299, 3300, 3301

Page 24
Greater than, less than

1. Answers will vary but should show:
   1085 m < 1344 m < 3970 m <
   4808 m < 8611 m < 8848 m
   and
   8848 m > 8611 m > 4808 m >
   3970 m > 1344 m > 1085 m
2. Snowdon, Ben Nevis, Eiger, Mont
   Blanc, K2, Mount Everest
Rocket Answers will vary.

Page 25
Rounding

1. $16 \rightarrow 20$
2. $23 \rightarrow 20$

3. 46 → 50
4. 43 → 40
5. 37 → 40
6. 51 → 50
7. 34 → 30
8. 22 → 20
9. 48 → 50
10. 64 → 60
11. 12 → 10
12. 73 → 70
13. 11 → 10
14. 82 → 80
15. 74 → 70

Rocket Answers will vary. An example might be numbers which round to 50: 45, 46, 47, 48, 49, 51, 52, 53 and 54.

Page 26
Rounding

1. 42 → 40
2. 86 → 90
3. 27 → 30
4. 95 → 100
5. 12 → 10
6. 54 → 50
7. 80 cm
8. 50 cm
9. 30 cm
10. 60 cm
11. 20 cm
12. 70 cm

Rocket 3 rounded up and 3 rounded down

Page 27
Rounding

1. 142 → 140
2. 64 → 60
3. 173 → 170
4. 273 → 270
5. 89 → 90
6. 385 → 390
7. 75 → 80
8. 496 → 500
9. 517 → 520
10. 134 g → 130 g
11. 217 g → 220 g
12. 121 g → 120 g
13. 142 g → 140 g
14. 175 g → 180 g
15. 182 g → 180 g
16. 187 g → 190 g
17. 133 g → 130 g
18. 158 g → 160 g

Rocket 4 weights rounded up and 5 rounded down

Page 28
Rounding

1. 61, 62, 64
2. 93, 94

---

3. 141, 142, 143, 144
4. 351, 352, 354
5. a: 33 → 30
   b: 36 → 40
   c: 39 → 40
6. d: 71 → 70
   e: 74 → 70
   f: 77 → 80
7. g: 162 → 160
   h: 165 → 170
   i: 166 → 170
   j: 168 → 170

Rocket 35, 36, 37, 38, 39, 41, 42, 43, 44

Page 29
Nearest 10, nearest 100

1. 125 → 100
2. 248 → 200
3. 569 → 600
4. 750 → 800
5. 933 → 900
6. 487 → 500

Rocket Answers will vary between 450 g and 500 g to 499 g and 549 g.

7. 468 → 470
   468 → 500
8. 217 → 220
   217 → 200
9. 652 → 650
   652 → 700
10. 712 → 710
    712 → 700
11. 902 → 900
    902 → 900
12. 395 → 400
    395 → 400
13. 555 → 560
    555 → 600
14. 313 → 310
    313 → 300

Page 30
Nearest 10, nearest 100

1. 143 km → 140 km
2. 189 km → 190 km
3. 155 km → 160 km
4. 139 km → 140 km
5. 134 km → 130 km

Towns in order of distance: Windy Village, Summer Town, Spring Park, Winter Gardens, Snowy Heights

6. 473 g → 470 g
   473 g → 500 g
7. 509 g → 510 g
   509 g → 500 g
8. 735 g → 740 g
   735 g → 700 g
9. 249 g → 250 g
   249 g → 200 g
10. 350 g → 350 g
    350 g → 400 g
11. 50 g → 50 g
    50 g → 100 g

---

12. 348 → 350, 300
13. 212 → 210, 200
14. 641 → 640, 600
15. 772 → 770, 800
16. 453 → 450, 500
17. 584 → 580, 600
18. 633 → 630, 600
19. 536 → 540, 500

Page 31
Nearest 10, nearest 100

1. 243 → 200
2. 483 → 500
3. 335 → 300
4. 451 → 500
5. 622 → 600
6. 395 → 400
7. 718 → 700
8. 690 → 700
9. 845 → 800
10. 54 → 50
11. 77 → 80
12. 98 → 100
13. 43 → 40
14. 340 → 300
15. 470 → 500
16. 120 → 100
17. 260 → 300

Rocket 195, 196, 197 and 198

Page 32
Nearest 10, nearest 100

1. c: 865, d: 891
2. a: 815
3. g: 1156
4. e: 1122, f: 1135
5. b: 846
6. e: 1122
7. i: 543
8. k: 579
9. n: 1387
10. j: 556
11. l: 1308, m: 1346
12. m: 1346

Rocket Answers will vary.

Number Processes PPMs

PPM 1
Dice numbers

1. 3
2. 2
3. 4
4. 1
5. 6
6. 5
7. 6
8. 8

PPM 2

Real life numbers

1. Answers will vary but should be a 2-digit number up to 20.
2. Answers will vary but should be an 11-digit number starting with 0.
3. Answers will vary but should be a 1-, 2- or 3-digit number.
4. Answers will vary but should be a 3- or 4-digit number.
5. Answers will vary but should be a 3- or 4-digit number.
6. Answers will vary depending on the size of your school.

PPM 3

Before and after

1. 4, 5, 6
2. 6, 7, 8
3. 2, 3, 4
4. 8, 9, 10
5. 5, 6, 7
6. 10, 11, 12
7. 1, 2, 3
8. 7, 8, 9
9. 14, 15, 16

PPM 4

Title: Between

1. 8, 9, 10
2. 4, 5, 6
3. 21, 22, 23
4. 12, 13, 14
5. 35, 36, 37
6. 14, 15, 16
7. 60, 61, 62
8. 52, 53, 54
9. 19, 20, 21
10. 38, 39, 40
11. 43, 44, 45

PPM 5

Missing numbers in a dial

1. 4, 6, 9
2. 5, 6, 7 … 9, 10, 11
3. 3, 4, 5
4. 51, 52, 53, 54

PPM 6

Queuing

Bee	6th
Stork	2nd
Bird	9th
Owl	4th
Frog	7th
Chick	1st
Cockerel	8th
Bear	11th

PPM 7

Position

1. 3rd — f, third
2. 9th — d, ninth
3. 15th — b, fifteenth
4. 21st — e, twenty-first
5. 18th — k, eighteenth
6. 5th — s, fifth
7. 26th — y, twenty-sixth
8. 11th — o, eleventh
9. 20th — n, twentieth
10. 12th — a, twelfth

PPM 8

Order in a race

	Position	In words	How many will finish in front?	How many will finish behind?
a	2nd	second	1	10
b	4th	fourth	3	8
c	6th	sixth	5	6
d	10th	tenth	9	2
e	12th	twelfth	11	0

PPM 9

2-digit numbers

1. 46
2. 37
3. 19
4. 63
5. 84
6. 91

PPM 10

More and less

	1 more	1 less	10 more	10 less
11	12	10	21	1
23	24	22	33	13
38	39	37	48	28
46	47	45	56	36
87	88	86	97	77
19	20	18	29	9
31	32	30	41	21
79	80	78	89	69
42	43	41	52	32
90	91	89	100	80

PPM 11

Number match

30 5	three 10s, five units	35
40 8	four 10s, eight units	48
6	no 10s, six units	6
60 6	six 10s, six units	66
40 1	four 10s, one units	41
10 4	one 10, four units	14

PPM 12

3-digit numbers

1. 185
2. 324
3. 264
4. 208
5. 640
6. 407

PPM 13

Hundreds, tens and units

1. 243
2. 256
3. 133
4. 222
5. 183
6. 360
7. 109
8. 404
9. 160

PPM 14

Hundreds, tens and units

1. 395 is 300 + 90 + 5
2. 456 is 400 + 50 + 6
3. 742 is 700 + 40 + 2
4. 707 is 700 + 0 + 7
5. 895 is 800 + 90 + 5
6. 414 is 400 + 10 + 4
7. 928 is 900 + 20 + 8
8. 458   4 hundreds, 5 tens, 8 units
9. 809   8 hundreds, 0 tens, 9 units
10. 1125 is 1000 + 100 + 20 + 5

PPM 15

1, 10, 100, 1000 more or less

1. 4647
2. 6315
3. 4078
4. 5832
5. 7407
6. 8389
7. 3154

8. 267
9. 5443
10. 6430

PPM 16

Largest, smallest and nearest number

1. 7432
2. 2358
3. 7641
4. 2483
5. 4721
6. 6843
7. 2953
8. 5126
9. 4873

PPM 17

Title: More or less

	1 more	10 less	100 more	1000 less
5836	5837	5826	5936	4836
2790	2791	2780	2890	1790
3025	3026	3015	3125	2025
4108	4109	4098	4208	3108
1893	1894	1883	1993	893
8947	8948	8937	9047	7947
5574	5575	5564	5674	4574
8049	8050	8039	8149	7049
7663	7664	7653	7763	6663
2951	2952	2941	3051	1951
3210	3211	3200	3310	2210
6499	6500	6489	6599	5499

18 numbers between 100 and 10 000
have all their digits the same:
111 one hundred and eleven
222 two hundred and twenty-two
333 three hundred and thirty three
444 four hundred and forty-four
555 five hundred and fifty-five
666 six hundred and sixty-six
777 seven hundred and seventy-seven
888 eight hundred and eighty-eight
999 nine hundred and ninety-nine
1111 one thousand, one hundred and
eleven
2222 two thousand, two hundred and
twenty-two
3333 three thousand, three hundred and
thirty three
4444 four thousand, four hundred and
forty-four
5555 five thousand, five hundred and
fifty-five
6666 six thousand, six hundred and
sixty-six
7777 seven thousand, seven hundred
and seventy-seven
8888 eight thousand, eight hundred and
eighty-eight
9999 nine thousand, nine hundred and
ninety-nine

PPM 18

Between

1. 15, 16, 17
2. 36, 37, 38
3. 54, 55, 56
4. 63, 64, 65
5. 70, 71, 72
6. 21, 22, 23
7. 98, 99, 100
8. 79, 80, 81

PPM 19

Caterpillars

1. 41, 40, 39, 38, 37, 36
2. 36, 35, 34, 33, 32, 31
3. 55, 54, 53, 52, 51, 50
4. 13, 12, 11, 10, 9, 8
5. 104, 103, 102, 101, 100, 99

PPM 20

Counting on and back in 10s

10 less	Start	10 more
25	35	45
34	44	54
50	60	70
2	12	22
69	79	89
73	83	93
11	21	31
80	90	100

PPM 21

10 and 100s towers

1. 120, 220, 320, 420, 520, 620, 720
2. 86, 186, 286, 386, 486, 586, 686
3. 35, 135, 235, 335, 435, 535, 635

PPM 22

Counting on and back in 100s

100 less	Start	100 more
400	500	600
250	350	450
675	775	875
746	846	946
122	222	322
92	192	292
308	408	508
3	103	203

PPM 23

Counting on and back in 10s

1. 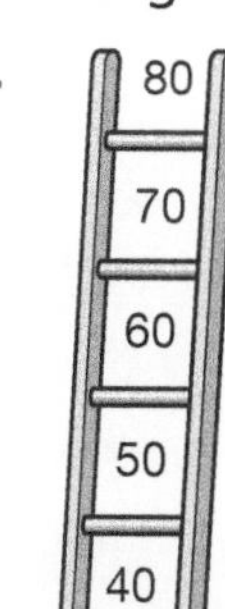

2. 

3. 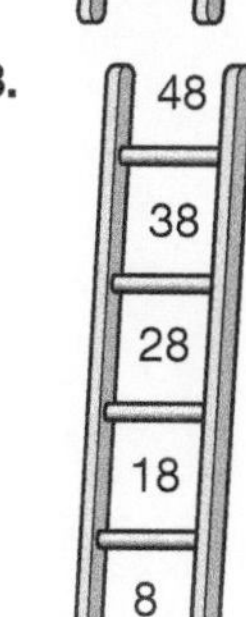

4. 

5. 

6. 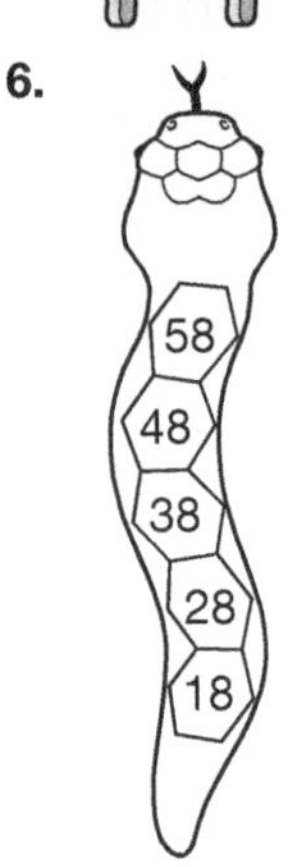

7. 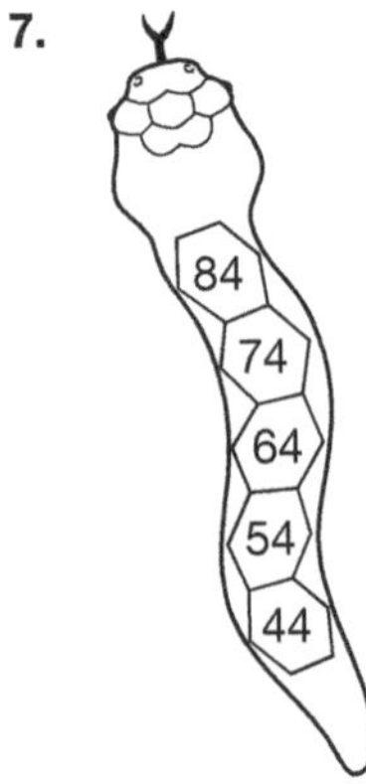

8. 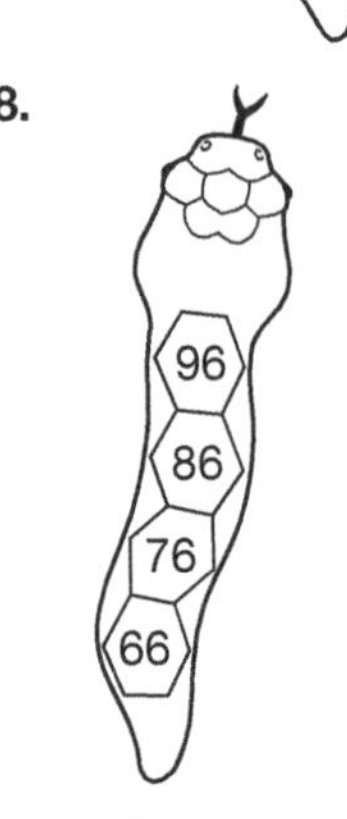

9. 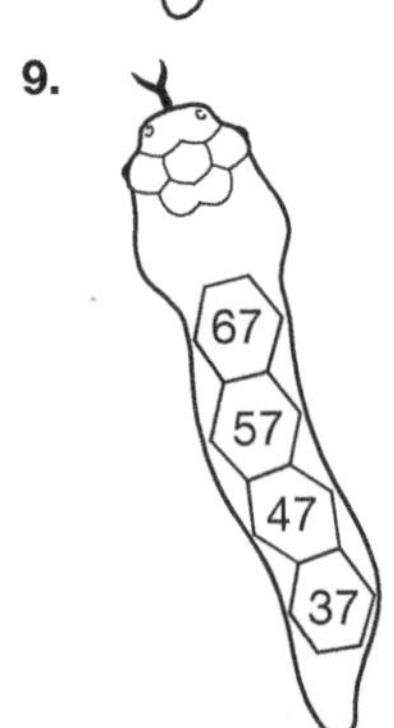

PPM 24

Numbers to 1000

1. 203, 206, 214, 217, 225, 228
2. 425, 426, 435, 439, 443, 447
3. 616, 620, 623, 628, 636, 639

PPM 25

Number words

6 letters
11 eleven
12 twelve
20 twenty
7 letters
15 fifteen
16 sixteen
8 letters
13 thirteen
14 fourteen
18 eighteen
19 nineteen
9 letters
17 seventeen

PPM 26

Counters

1. 22
2. 34
3. 19
4. 26
5. 13
6. 45
7. Children draw two columns of 10 counters and five units counters
8. Children draw three columns of 10 counters and six units counters
9. Children draw four columns of 10 counters and one units counter

PPM 27

Matching 2-digit number words

fifteen	15
twenty-four	24
forty-seven	47
sixty-three	63
14	fourteen
98	ninety-eight

PPM 28

Writing 3-digit numbers

156	one hundred and fifty-six
621	six hundred and twenty-one
235	two hundred and thirty-five
241	two hundred and forty one
612	six hundred and twelve
905	nine hundred and five
872	eight hundred and seventy-two
404	four hundred and four

PPM 29

Matching 3-digit number words

two hundred and forty-six	246
four hundred and twelve	412
seven hundred and ninety-five	795
three hundred and sixteen	316
eight hundred and six	806
nine hundred and ninety-eight	998

PPM 30

4-digit numbers

This is a game.

PPM 31

5-digit numbers

This is a game.

PPM 32

Smallest to largest

1. 3, 4, 5
2. 1, 2, 3
3. 4, 5, 6
4. 8, 9, 10
5. 9, 10, 11, 12
6. 17, 20, 23, 24
7. 29, 31, 33, 37
8. 40, 46, 52, 57
9. 83, 85, 87, 89

PPM 33

Ordering

1. 9, 13, 15, 17, 18
2. 11, 16, 21, 28, 32
3. 25, 33, 42, 51, 64
4. 35, 39, 42, 47, 56
5. 73, 79, 80, 82, 91

PPM 34

Ordering

1. 35, 46, 86, 93, 125
2. 18, 29, 76, 95, 128
3. 9, 32, 56, 117, 141
4. 38, 46, 57, 99, 213
5. 16, 18, 52, 98, 325

PPM 35

Card numbers

25, 27, 29, 52, 57, 59, 72, 75, 79, 92, 95, 97

PPM 36

Ordering in hundreds

1. 85, 97, 102, 111, 120
2. 256, 348, 417, 526, 634
3. 408, 452, 463, 471, 479
4. 320, 329, 330, 331, 333
5. 757, 758, 759, 760, 761

PPM 37

Ordering in thousands

1. 534, 542, 543, 561
2. 625, 921, 4236, 7624
3. 1234, 4370, 5545, 6085
4. 3689, 4399, 4420, 6127
5. 2260, 2282, 2317, 2428

PPM 38

More than, less than, in between

Answers will vary for the missing digits exercise on the first half of PPM 38.

1. >, 75
2. <, 33 (for questions 2–9 numerical answers may vary)
3. <, 95
4. >, 58
5. >, 35
6. >, 63
7. >, 22
8. >, 94
9. <, 107

PPM 39

Equals

1. $26 + 1 = 28 - 1$
2. $34 + 2 \neq 30 + 5$
3. $63 + 2 = 67 - 2$
4. $38 - 5 \neq 36 - 2$
5. $43 + 2 \neq 48 - 5$
6. $95 - 5 \neq 86 + 5$
7. Answers will vary but should not be $72 + 5 \neq 79 - 2$
8. Answers will vary but should not be $23 + 5 \neq 32 - 4$
9. Answers will vary but should not be $84 + 2 \neq 88 - 2$
10. Answers will vary but should not be equal number sentences.

11–14. Children's own examples will vary.

PPM 40

Greater than, less than, equals

1. $25 < 33 > 17 = 17 < 38 > 21$
2. Answers will vary, for example:
   $32 < 54 > 9 = 9 < 72 < 80$
3. $42 < 78 = 78 < 92 > 47 <$ any number less than 47

4–5. Children's own examples will vary.

6. Answers will vary, for example:
   $9 + 5 > 16 - 3 = 7 + 6$
7. Answers will vary, for example:
   $23 + 9 > 34 - 5 = 12 + 17$
8. Answers will vary, for example:
   $18 + 5 < 69 - 5 > 51 + 7$

PPM 41

Greater than, less than

1. <
2. <
3. <
4. <
5. <
6. >
7. <
8. >
9. >
10. <
11. >
12. <
13. >
14. >
15. <
16. >
17. >
18. >

PPM 42

Greater than less than

1. 3536, 3546, 3556, 3566, 3576, 3586 or 3596
2. 8104, 8114, 8124, 8134, 8144, 8154 or 8164
3. 8392
4. 5670, 5671, 5672, 5673, 5674, 5675 or 5676
5. 3529, 3629, 3729, 3829 or 3929
6. 1153, 2153, 3153, 4153, 5153 or 6153
7. 2705, 2715, 2725, 2735, 2745, 2755, 2765, 2775, 2785 or 2795
8. 3999
9. 7265, 7275, 7285 or 7295
10. 8650, 8651, 8652, 8653, 8654, 8655, 8656, 8657, 8658 or 8659
11. 2740, 2750, 2760, 2770, 2780 or 2790
12. 3790
13. 5406v
14. 9309, 9319, 9329, 9339, 9349, 9359, 9369 or 9379
15. 47 320, 47 321, 47 322, 47 323, 47 324 or 47 325
16. 51 406 or 51 416

PPM 43

Equal or not equal

1. $328 + 20 = 358 - 10$
2. $446 + 30 \neq 496 - 30$
3. $633 + 20 = 673 - 20$
4. $384 - 50 \neq 364 - 40$
5. $439 + 40 = 499 - 20$
6. $795 - 50 \neq 755 + 40$
7. Answers will vary, for example:
   $719 + 50 \neq 789 - 40$
8. Answers will vary, for example:
   $232 + 40 \neq 292 - 60$
9. Answers will vary, for example:
   $814 + 50 \neq 874 - 20$
10. Answers will vary, for example:
    $308 - 60 \neq 278 + 10$

11–14. Children's own examples will vary.

PPM 44

Rounding to the nearest 10

1. Any answers from 15, 16, 17, 18, 19, 21, 22, 23, 24
2. Any answers from 35, 36, 37, 38, 39, 41, 42, 43, 44
3. Any answers from 55, 56, 57, 58, 59, 61, 62, 63, 64
4. Any answers from 75, 76, 77, 78, 79, 81, 82, 83, 84
5. Any answers from 5, 6, 7, 8, 9, 11, 12, 13, 14
6. Any answers from 95, 96, 97, 98, 99, 101, 102, 103, 104

PPM 45

Rounding

1. position   74 78 83 86 89
   nearest 10   70 80 80 90 90
2. position   203 207 211 215 218
   nearest 10   200 210 210 220 220
3. position   128 134 144 156 168
   nearest 10   130 130 140 160 170

PPM 46

Rounding in calculations

1. 60
2. 20
3. 60
4. 50
5. 20
6. 20
7. 10
8. 30
9. 60

PPM 47

Nearest 100

Answers should include any twelve from:

$247 \rightarrow 200$	$724 \rightarrow 700$
$249 \rightarrow 200$	$729 \rightarrow 700$
$274 \rightarrow 300$	$742 \rightarrow 700$
$279 \rightarrow 300$	$749 \rightarrow 700$
$294 \rightarrow 300$	$792 \rightarrow 800$
$297 \rightarrow 300$	$794 \rightarrow 800$
$427 \rightarrow 400$	$924 \rightarrow 900$
$429 \rightarrow 400$	$927 \rightarrow 900$
$472 \rightarrow 500$	$942 \rightarrow 900$
$479 \rightarrow 500$	$947 \rightarrow 900$
$492 \rightarrow 500$	$972 \rightarrow 1000$
$497 \rightarrow 500$	$974 \rightarrow 1000$

PPM 48

Rounding to the nearest 100

1. Any answers from 151 to 249
2. Any answers from 351 to 449
3. Any answers from 551 to 649
4. Any answers from 751 to 849
5. Any answers from 51 to 149
6. Any answers from 951 to 1049

Number Processes APMs

APM 92

Race to the top!

Race 1: finishing order is F, J, D, I, B, H, A, G, E, C

Race 2: finishing order is: G, E, A, I, J, C, F, B, H, D

Addition and Subtraction Pupil Book 2

Page 3

Counting on in 11s

1. 2, 13, 24, 35, 46, 57, 68, 79, 90
2. 3, 14, 25, 36, 47, 58, 69, 80, 91
3. 6, 17, 28, 39, 50, 61, 72, 83, 94
4. 8, 19, 30, 41, 52, 63, 74, 85, 96
5. 12, 23, 34, 45, 56, 67, 78, 89, 100
6. 16, 27, 38, 49, 60, 71, 82, 93, 104

Rocket

Number	Digit total
2	2
13	4
24	6
35	8
46	10
57	12
68	14
79	16
90	9
101	2
112	4
123	6
134	8
145	10
156	12
167	14
178	16
189	18
200	2
211	4

Children will notice that most of the totals form part of the 2-times table.

7. 6 pies.
8. 9 weeks.

Page 4

Counting in 50s

1. 150
2. 300
3. 400
4. 250
5. 350
6. 200
7. 450
8. 650
9.

Number of bags	10	15	13	18	7	22	17	39
Number of tacks	500	750	650	900	350	1100	850	1950

10. 442 g

Rocket About 38 or 39 years

Page 5

Counting in 25s and 50s

1.

	600	705	725	1005	810	100	48	130	350	950	425
In 50s count	✓					✓			✓	✓	
In 25s count	✓		✓			✓			✓	✓	✓

2. The pattern is that every alternate number increases by 100 each time.

Rocket

Number	25	50	75	100	125	150	175	200
Digit total	7	5	3	1	8	6	4	2

The pattern is that the digit totals go down in steps of 2 to the smallest possible number, first odd numbers then even numbers.

3. 75, 100, 125
4. 225, 250, 275
5. 350, 375, 400
6. 150, 175, 200
7. 125, 150, 175
8. 275, 300, 325
9. 14 weeks
10. 40 days

Page 6

Counting in 25s

1. 1300 g or 1·3 kg

Rocket 10th number: 257
100th number: 2507

2. 59 cm, 34 cm, 9 cm
3. 43 cm, 18 cm
4. 70 cm, 45 cm, 20 cm
5. 31 cm, 6 cm
6. 40 cm, 15 cm
7. 54 cm, 29 cm, 4 cm

Page 7

Addition and subtraction facts

1. Answers will vary. For example
   $7 - 3 = 4$
   $30 + 40 = 7$
   $23 + 24 = 47$
2. The easier calculation is $2 + 5$.
   $2 + 5 = 7$ so $200 + 500 = 700$
3. $8 - 5 = 3$
   $18 - 5 = 13$
   $28 - 5 = 23$
   $38 - 5 = 33$
   $48 - 5 = 43$
   $58 - 5 = 53$
4. $3 + 5 = 8$
   $13 + 5 = 18$
   $23 + 5 = 28$
   $33 + 5 = 38$
   $43 + 5 = 48$
   $53 + 5 = 58$

Rocket Answers will vary, for example
$13 - 7 = 6$ $13 - 6 = 7$
$106 + 107 = 213$ $60 + 70 = 130$
$130 - 60 = 70$ $130 - 70 = 60$
$16 + 17 = 33$

Page 8

Number pairs to 100

1. 36
2. 28
3. 46
4. 62
5. 43
6. 31
7. 48
8. 33
9. 65

Rocket Answers will vary. Possible answers include: 20, 80; 30, 70. The rule is they are multiples of 10.

10. 66
11. 73
12. 64p, 28p
13. 53

Page 9

Adding

1. $30 + 20 = 50$, $6 + 3 = 9$,
   $50 + 9 = 59$
2. $50 + 20 = 70$, $4 + 4 = 8$,
   $70 + 8 = 78$
3. $60 + 20 = 80$, $7 + 1 = 8$,
   $80 + 8 = 88$
4. $70 + 20 = 90$, $2 + 7 = 9$,
   $90 + 9 = 99$
5. $50 + 40 = 90$, $3 + 4 = 7$,
   $90 + 7 = 97$
6. $60 + 20 = 80$, $5 + 5 = 10$,
   $80 + 10 = 90$
7. $40 + 30 = 70$, $8 + 6 = 14$,
   $70 + 14 = 84$
8. $50 + 30 = 80$, $5 + 7 = 12$,
   $80 + 12 = 92$

9.   50 + 30 = 80, 6 + 6 = 12,
      80 + 12 = 92
10. 30 + 40 = 70, 9 + 4 = 13,
      70 + 13 = 83
11. 50 + 20 = 70, 7 + 6 = 13,
      70 + 13 = 83
12. 40 + 30 = 70, 2 + 8 = 10,
      70 + 10 = 80
13. £8
Rocket Answers will vary.

Page 10
Subtracting multiples of 10

1.   86 – 50 = 36
2.   75 – 20 = 55
3.   92 – 30 = 62
4.   66 – 40 = 26
5.   85 – 60 = 25
6.   71 – 50 = 21
7.   36 – 10 = 26
8.   47 – 20 = 27
9.   78 – 40 = 38
10. 55 – 40 = 15
Rocket Answers will vary.
11. 44 marbles.
12. 12 football cards.
13. 42 pennies.

Page 11
Addition and subtraction facts

1.   50 – 18 = 32
2.   70 – 18 = 52
3.   67 – 43 = 24
4.   53 – 24 = 29
5.   28 = 74 – 46
6.   45 = 78 – 33
7.   51 = 96 – 45
8.   62 = 91 – 29
9.   74 = 99 – 25
10. 18 = 46 – 28
11. 14 = 78 – 64
12. 75 = 101 – 26
13. 16 marbles.
14. 28 books.

Page 12
Adding and subtracting

1.   27 + 7 = 34
2.   84 + 8 = 92
3.   47 + 6 = 53
4.   55 – 8 = 47
5.   54 + 8 = 62
6.   65 – 7 = 58
7.   73 + 8 = 81
8.   44 – 5 = 39
9.   53 – 5 = 48
10. 46 + 7 = 53
11. 37 + 6 = 43
12. 52 – 8 = 44
13. 38 – 4 = 34
14. 45 + 6 = 51

Rocket This is only true when the other
      number is even, for example
      2 + 2 = 4.

Page 13
Adding 2-digit numbers

1.   47 + 36 = 70 + 13 = 83
2.   52 + 48 = 90 + 10 = 100
3.   23 + 49 = 60 + 12 = 72
4.   78 + 17 = 80 + 15 = 95
5.   63 + 48 = 100 + 11 = 111
6.   36 + 27 = 50 + 13 = 63
7.   53 + 66 = 110 + 9 = 119
8.   45 + 74 = 110 + 9 = 119
9.   33 + 57 = 80 + 10 = 90
10. 48p + 67p = 115p
11. 74p + 65p = 139p
12. 81p + 49p = 130p
13. 56p + 72p = 128p
14. 47p + 63p = 110p
15. 63p + 59p = 122p
Rocket 22 + 99, 33 + 88, 44 + 77, 55 + 66

Page 14
Adding several numbers

1.   8 + 2 = 10, 10 + 6 = 16, 16 + 9 = 25
2.   3 + 7 = 10, 10 + 8 = 18, 18 + 8 = 26
3.   4 + 6 = 10, 10 + 9 = 19, 19 + 6 = 25
4.   9 + 1 = 10, 10 + 9 = 19, 19 + 4 = 23
5.   3 + 7 = 10, 10 + 5 = 15, 15 + 6 = 21
6.   5 + 5 = 10, 10 + 4 = 14, 14 + 3 = 17
Rocket Answers will vary. An example
      might be 6, 6, 8 and 5.
7.   11p + 5p + 9p + 3p = 28p
8.   12p + 8p + 5p + 2p = 27p
9.   2p + 14p 8p + 6p = 30p
10. 5p + 15p + 6p + 8p = 34p
11. 3p + 6p + 4p + 17p = 30p
12. 3p + 16p + 8p + 5p = 32p

Page 15
Adding several numbers

1.   8 + 2 + 5 + 3 = 18 minutes
2.   6 + 4 + 8 + 7 = 25 minutes
3.   5 + 5 + 9 + 9 + 6 = 34 minutes
4.   3 + 7 + 3 + 4 + 9 = 26 minutes
Rocket Answers will vary.
5.   30 + 70 + 80 + 40 = 220
6.   50 + 50 + 60 + 80 = 240
7.   60 + 40 + 70 + 70 = 240
8.   80 + 20 + 70 + 30 + 40 = 240
9.   90 + 10 + 70 + 90 = 260
10. 70 + 30 + 60 + 90 = 250
11. 80 + 20 + 60 + 90 = 250
12. 60 + 40 + 80 + 50 = 230

Page 16
Adding several numbers

1.   22p + 7p + 3p + 9p = 41p
2.   2p + 12p + 9p + 35p + 8p = 66p
3.   26p + 8p + 9p + 1p + 4p = 48p

4.   6p + 4p + 11p + 7p + 5p = 33p
5.   24 + 9 + 8 + 4 + 3 = 48; 2 more to
      make 50 marbles
6.   £46 + £18 + £6 + £14 = £84
7.   Answers will vary. An example
      might be 4p + 6p + 8p + 2p + 19p
      = 39p.
Rocket Any number between 1p and
      50p, that is a multiple of 5 (i.e.
      5p, 10p, 15p …)

Page 17
Adding several numbers

1.   22p + 8p + 3p + 7p + 9p = 49p
2.   9p + 17p + 6p + 4p + 8p = 44p
3.   17p + 5p + 24p + 19p + 3p = 68p
4.   2p + 25p + 9p + 8p + 24p = 68p
5.   4p + 27p + 3p + 16p + 4p = 54p
6.   7p + 21p + 19p + 14p + 2p = 63p
7.   25p + 8p + 4p + 25p + 9p = 71p
8.   27p + 8p + 4p + 7p + 3p + 5p = 54p
9.   6p + 27p + 4p + 9p + 2p = 48p
10. 24p + 9p + 6p + 17p + 3p = 59p
11. 18p + 9p + 8p + 4p + 23p = 62p
12. 5p + 8p + 23p + 18p + 4p = 58p
13. Answers will vary. If children use
      stamps with values of 1p, 2p, 5p,
      10p and 20p the answers will be:
      5, 4, 6, 6, 5, 5, 5, 5, 5, 6, 4, 6.

Page 18
Adding and subtracting

1.   56 + 9 = 65
2.   27 + 9 = 36
3.   38 + 9 = 47
4.   65 + 9 = 74
5.   73 + 9 = 82
6.   46 + 9 = 55
7.   27 + 11 = 38
8.   43 + 11 = 54
9.   28 + 11 = 39
10. 78p – 9p = 69p
11. 65p – 9p = 56p
12. 87p – 9p = 78p
13. 56p – 9p = 47p
14. 42p – 9p = 33p
15. 64p – 9p = 55p
16. 92p – 9p = 83p
17. 85p – 9p = 76p
18. 54p – 9p = 45p
19. 43p – 9p = 34p
Each price is increased by 9p:
10. 78p + 9p = 87p
11. 65p + 9p = 74p
12. 87p + 9p = 96p
13. 56p + 9p = 65p
14. 42p + 9p = 51p
15. 64p + 9p = 73p
16. 92p + 9p = 101p
17. 85p + 9p = 94p
18. 54p + 9p = 63p
19. 43p + 9p = 52p
Rocket 11 times

Page 19

Adding and Subtracting

1. 24 + 19 = 43
2. 42 + 19 = 61
3. 48 + 19 = 67
4. 63 + 21 = 84
5. 36 + 21 = 57
6. 72 + 21 = 93
7. False
8. False
9. True
10. £46 – £19 = £27
11. £42 – £19 = £23
12. £38 – £19 = £19
13. £56 – £21 = £35
14. £64 – £21 = £43
15. £88 – £21 = £67

Rocket When adding 19, the units digit is 1 less than in the first number. When subtracting 19, the units digit is 1 more than in the first number.

Page 20

Adding and Subtracting

1. 354 + 39 = 393
2. 327 + 49 = 376
3. 312 + 29 = 341
4. 332 + 29 = 361
5. 325 + 59 = 384
6. 443 + 39 = 482
7. 250 – 29 = 221
   250 – 41 = 209
   250 – 99 = 151
   250 – 19 = 231
   250 – 39 = 211
   250 – 59 = 191
   250 – 119 = 131
   405 fish are not eaten altogether.
8. £582 – £39 = £543
9. £476 – £29 = £447
10. £553 – £31 = £522
11. £397 – £51 = £346

Rocket 16 times, with 56 left.

Page 21

Adding

1. 26 + 23 = 59
2. 36 + 42 = 78
3. 42 + 33 = 75
4. 45 + 34 = 79
5. 17 + 42 = 59
6. 32 + 12 = 44
7. 51 + 25 = 76
8. 28 + 91 = 119
9. Answers will vary.

Rocket 25 + 26 = 51

Page 22

Adding

1. 374 + 268 = 642
2. 286 + 418 = 704
3. 375 + 188 = 563
4. 278 + 187 = 465
5. 392 + 129 = 521
6. 188 + 276 = 464
7. 283 + 379 = 662
8. 235 + 566 = 801
9. 294 + 516 = 810
10. 1900 + 28 + 46 + 27 = 2001.
11. 1898 + 28 + 25 + 28 = 1999

Page 23

Subtracting

1. 82 – 18 = 64 cm
2. 73 – 36 = 37 cm
3. 94 – 25 = 69 cm
4. 65 – 28 = 37 cm
5. 71 – 27 = 44 cm
6. 63 – 15 = 48 cm
7. 208 – 32 = 176 cm
8. 183 – 58 = 125 cm
9. 274 – 37 = 237 cm

Rocket Answers will vary. For example, a bush 80 cm high would have 40 cm cut off to make it half the height, then 20 cm to make it half the height again, then 10 cm, then 5 cm then 2.5 cm, etc

10. 183 – 47 = 136
11. 221 – 188 = 33
12. 144 – 86 = 58
13. 164 – 55 = 109
14. 214 – 187 = 27
15. 372 – 136 = 236

Page 24

Adding and subtracting multiples of 10

1. 14 – 6 = 8
   140 – 60 = 80
2. 17 + 5 = 22
   170 + 50 = 220
3. 16 – 7 = 9
   160 – 70 = 90
4. 24 + 8 = 32
   240 + 80 = 320
5. 12 – 5 = 7
   120 – 50 = 70
6. 17 + 9 = 26
   170 + 90 = 260
7. 18 + 7 = 25
   180 + 70 = 250
8. 13 – 7 = 6
   130 – 70 = 60
9. 15 – 8 = 7
   150 – 80 = 70
10. 23 – 9 = 14
    230 – 90 = 140
11. 21 – 8 = 13
    210 – 80 = 130
12. 11 – 7 = 4
    110 – 70 = 40

13. Answers will vary.

Rocket Red cards: 150 + 320 + 240 + 510 + 120 = 1340. Yellow cards: 50 + 90 + 70 + 60 = 270. Red – Yellow: 1340 – 270 = 1070

Page 25

Adding and subtracting multiples of 10

1. Total: 310 + 70 = 380
   Difference: 310 – 70 = 240
2. Total: 220 + 70 = 290
   Difference: 220 – 70 = 150
3. Total: 440 + 80 = 520
   Difference: 440 – 80 = 360
4. Total: 230 + 90 = 320
   Difference: 230 – 90 = 140
5. Total: 340 + 60 = 400
   Difference: 340 – 60 = 280
6. Total: 150 + 80 = 230
   Difference: 150 – 80 = 70
7. Total: 210 + 80 = 290
   Difference: 210 – 80 = 130

Rocket
1. £3100 + £350 = £3450
2. £2200 + £350 = £2550
3. £4400 + £400 = £4800
4. £2300 + £450 = £2750
5. £3400 + £300 = £3700
6. £1500 + £400 = £1900
7. £2100 + £400 = £2500
8. 320 – 60 = 260
9. 430 – 80 = 350
10. 370 + 70 = 440
11. 360 + 80 = 440
12. 880 + 40 = 920
13. 830 – 60 = 770
14. 710 – 90 = 620
15. 280 + 50 = 330
16. 640 + 50 = 690

Page 26

Adding and subtracting multiples of 10

1. 446 + 50 = 496 km/h
2. 787 + 50 = 837 km/h
3. 367 + 30 = 397 km/h
4. 472 + 40 = 512 km/h
5. 685 + 50 = 735 km/h
6. 856 + 60 = 916 km/h
7. 779 + 40 = 819 km/h
8. 566 + 60 = 626 km/h
9. 832 – 50 = 782 m
10. 714 – 60 = 654 m
11. 653 – 60 = 593 m
12. 524 – 70 = 454 m
13. 333 – 60 = 273 m
14. 426 – 40 = 386 m
15. 505 – 50 = 455 m
16. 843 – 60 = 783 m

Rocket 12 and a half seconds

Page 27
Multiples of 100

1. 8400 – 300 = 8100 m
2. 7800 – 400 = 7400 m
3. 7500 – 500 = 7000 m
4. 6900 – 500 = 6400 m
5. 6400 – 200 = 6200 m
6. 5800 – 300 = 5500 m
7. £2700 + £500 = £3200
8. £4500 + £600 = £5100
9. £3600 + £800 = £4400
10. £5800 + £500 = £6300

Rocket Answers will vary. There are numerous possible ways for the Rocket to get to the ground.

Page 28
Number pairs

1.
```
    5 9 p
  + 2 7
    8 6 p
      1
```
2.
```
    6 8 p
  + 1 8
    8 6 p
      1
```
3.
```
    6 7 p
  + 2 6
    9 3 p
      1
```
4.
```
    3 5 p
  + 5 7
    9 2 p
      1
```
5.
```
    3 3 p
  + 4 4
    7 7 p
```
6.
```
    3 5 p
  +   6 7
    1 0 2 p
        1
```
7.
```
    3 6 p
  +   7 7
    1 1 3 p
        1
```
8.
```
    5 6 p
  + 2 7
    8 3 p
      1
```
9.
```
    4 2 p
  +   8 9
    1 3 1 p
        1
```
10.
```
    3 5 p
  +   9 7
    1 3 2 p
        1
```
11.
```
    7 3 p
  +   4 7
    1 2 0 p
        1
```
12.
```
    5 6 p
  +   5 6
    1 1 2 p
        1
```

Rocket Answers will vary.

Page 29
Adding

1.
```
  H T U
  1 2 7
      5 4
  + 1 1 5
  2 9 6 kg
      1
```
2.
```
  H T U
  1 2 8
  1 1 0
  +   3 6
  2 7 4 kg
      1
```
3.
```
  H T U
  1 1 2
      7 3
  +   2 3
  2 0 8 kg
      1
```
4.
```
  H T U
  1 0 9
      5 7
  +   8 5
  2 5 1 kg
    1 2
```

Rocket 12 + 345 + 6789 = 7146
5. 301 miles
6. 225 days
Rocket Answers will vary.

Page 30
Adding

Children should also include estimates.

1.
```
  H T U
  4 3 7
  + 1 4 6
  5 8 3 g
```
2.
```
  H T U
  3 7 8
  + 1 6 6
  5 4 4 g
```
3.
```
  H T U
  5 2 6
  + 2 1 9
  7 4 5 g
```
4.
```
  H T U
  3 5 9
  + 1 7 6
  5 3 5 g
```
5.
```
  H T U
  4 6 8
  + 1 2 9
  5 9 7 g
```
6.
```
  H T U
  3 3 6
  + 1 5 8
  4 9 4 g
```
7. Answers will vary.
Rocket Answers will vary. Possible answers include: 176 + 324 = 500, 174 + 326 = 500.

Page 31
Adding

1.
Calculation	Rounding
1 6 7	1 7 0
7 2	7 0
+ 8 6	+ 9 0
3 2 5	3 3 0
2 1	2

2.
Calculation	Rounding
1 5 4	1 5 0
6 5	7 0
+ 2 3	+ 2 0
2 4 2	2 4 0
1 1	1

3.
Calculation	Rounding
1 7 5	1 8 0
5 4	5 0
+ 3 8	+ 4 0
2 6 7	2 7 0
1 1	1

4.
Calculation	Rounding
1 6 5	1 7 0
3 8	4 0
+ 4 6	+ 5 0
2 4 9	2 6 0
1 1	1

Rocket 74 + 63 + 52 = 189 (or other combinations involving 70, 60, 50 and 2, 3, 4)
5. 579 miles
6. 450 days. Today's date is 26th March (as long as it isn't a leap year, in which case today's date will be 25th March).
7. 1328 g or 1 kg 328 g

Page 32
Adding

1. Answers will vary.
Rocket
£555 + £467 + £459 = £1481
£555 + £467 + £386 = £1408
£555 + £459 + £386 = £1400
£555 + £467 + £368 = £1390
£555 + £459 + £368 = £1382
£467 + £459 + £386 = £1312
£555 + £386 + £368 = £1309
£467 + £459 + £368 = £1294
£467 + £386 + £368 = £1221
£459 + £386 + £368 = £1213

Page 33
Subtracting

Children should also include estimates.

1.
```
  6 7 13
  -   4 7
      2 6
```
2.
```
  5 6 11
  -   4 3
      1 8
```
3.
```
  7 8 12
  -   5 8
      2 4
```

4.

$$
\begin{array}{r}
{}^{6}\cancel{7}\,{}^{1}1 \\
-\ 3\ 8 \\
\hline
3\ 3
\end{array}
$$

5.

$$
\begin{array}{r}
{}^{5}\cancel{6}\,{}^{1}2 \\
-\ 2\ 7 \\
\hline
3\ 5
\end{array}
$$

6.

$$
\begin{array}{r}
{}^{6}\cancel{7}\,{}^{1}4 \\
-\ 4\ 6 \\
\hline
2\ 8
\end{array}
$$

Rocket Answers will vary. An example
might be: 13 and 26;
26 − 13 = 13. The larger number
is double the smaller number.
The larger number is always
even.

7. In any order, six of:
82 − 74 = 8
82 − 65 = 17
82 − 57 = 25
82 − 38 = 44
74 − 65 = 9
74 − 57 = 17
74 − 38 = 36
65 − 57 = 8
65 − 38 = 27
57 − 38 = 19

Page 34
Subtracting

1.

$$
\begin{array}{r}
{}^{2}\cancel{3}\,{}^{1}4\ \ 6 \\
-\quad\ 8\ \ 4 \\
\hline
2\ \ 6\ \ 2
\end{array}
$$

2.

$$
\begin{array}{r}
{}^{1}2\,{}^{1}4\ \ 8 \\
-\quad\ 7\ \ 3 \\
\hline
1\ \ 7\ \ 5
\end{array}
$$

3.

$$
\begin{array}{r}
\cancel{7}\,{}^{1}2\ \ 6 \\
-\quad\ 6\ \ 2 \\
\hline
6\ \ 4
\end{array}
$$

4.

$$
\begin{array}{r}
4\ {}^{7}\cancel{8}\,{}^{1}2 \\
-\quad\ 7\ \ 6 \\
\hline
4\ \ 0\ \ 6
\end{array}
$$

5.

$$
\begin{array}{r}
{}^{2}\cancel{3}\,{}^{1}3\ \ 5 \\
-\quad\ 8\ \ 2 \\
\hline
2\ \ 5\ \ 3
\end{array}
$$

6.

$$
\begin{array}{r}
\cancel{7}\,{}^{1}3\ \ 7 \\
-\quad\ 8\ \ 5 \\
\hline
5\ \ 2
\end{array}
$$

7.

$$
\begin{array}{r}
{}^{1}2\,{}^{1}2\ \ 7 \\
-\quad\ 6\ \ 4 \\
\hline
1\ \ 6\ \ 3
\end{array}
$$

8.

$$
\begin{array}{r}
\cancel{7}\,{}^{1}2\,{}^{1}4 \\
-\quad\ 8\ \ 6 \\
\hline
3\ \ 8
\end{array}
$$

9. 115 pages
10. 214 pages
11. 316 pages
12. 427 pages
13. 345 pages
14. 436 pages
Rocket Answers will vary.

Page 35
Subtracting

1.

$$
\begin{array}{r}
{}^{2}\cancel{3}\,{}^{14}\cancel{5}\ \ 4 \\
-\ 1\ \ 6\ \ 7 \\
\hline
1\ \ 8\ \ 7
\end{array}
$$

2.

$$
\begin{array}{r}
{}^{3}\cancel{4}\,{}^{11}2\ \ {}^{1}3 \\
-\ 1\ \ 7\ \ 6 \\
\hline
2\ \ 4\ \ 7
\end{array}
$$

3.

$$
\begin{array}{r}
{}^{4}\cancel{5}\,{}^{11}2\ \ {}^{1}4 \\
-\ 1\ \ 7\ \ 8 \\
\hline
3\ \ 4\ \ 6
\end{array}
$$

4.

$$
\begin{array}{r}
{}^{2}\cancel{3}\,{}^{11}2\ \ {}^{1}5 \\
-\ 1\ \ 5\ \ 8 \\
\hline
1\ \ 6\ \ 7
\end{array}
$$

5.

$$
\begin{array}{r}
{}^{3}\cancel{4}\,{}^{12}3\ \ {}^{1}6 \\
-\ 1\ \ 8\ \ 7 \\
\hline
2\ \ 4\ \ 9
\end{array}
$$

6.

$$
\begin{array}{r}
{}^{2}\cancel{3}\,{}^{13}\cancel{4}\ \ {}^{1}1 \\
-\ 1\ \ 7\ \ 4 \\
\hline
1\ \ 6\ \ 7
\end{array}
$$

7. Correct
8. Miscalculated 14 − 9, should be:

$$
\begin{array}{r}
2\ {}^{7}\cancel{8}\,{}^{1}4 \\
-\ 1\ \ 3\ \ 9 \\
\hline
1\ \ 4\ \ 5
\end{array}
$$

9. No need to exchange for 6 − 3,
should be:

$$
\begin{array}{r}
{}^{3}\cancel{4}\,{}^{1}1\ \ 6 \\
-\ 1\ \ 8\ \ 3 \\
\hline
2\ \ 3\ \ 3
\end{array}
$$

10. Correct
11. Need to exchange for 20 − 80,
should be:

$$
\begin{array}{r}
{}^{4}\cancel{5}\,{}^{12}\ \ 4 \\
-\ 1\ \ 8\ \ 3 \\
\hline
3\ \ 4\ \ 1
\end{array}
$$

12. Correct
Rocket Largest answer: 875 − 134 = 741
Smallest answer: 513 − 487 = 26
Closest to 100: 481 − 375 = 106
475 − 381 = 94
841 − 735 = 106
835 − 741 = 94

Page 36
Subtracting

1.

$$
\begin{array}{r}
{}^{6}\cancel{7}\,{}^{10}\cancel{1}\ \ {}^{1}2 \\
-\ 3\ \ 4\ \ 5 \\
\hline
3\ \ 6\ \ 7
\end{array}
$$

2.

$$
\begin{array}{r}
{}^{6}\cancel{7}\,{}^{10}\cancel{1}\ \ {}^{1}2 \\
-\ 1\ \ 7\ \ 9 \\
\hline
5\ \ 3\ \ 3
\end{array}
$$

3.

$$
\begin{array}{r}
{}^{6}\cancel{7}\,{}^{10}\cancel{1}\ \ {}^{1}2 \\
-\ 2\ \ 6\ \ 8 \\
\hline
4\ \ 4\ \ 4
\end{array}
$$

4.

$$
\begin{array}{r}
{}^{6}\cancel{7}\,{}^{10}\cancel{1}\ \ {}^{1}2 \\
-\ 5\ \ 3\ \ 4 \\
\hline
1\ \ 7\ \ 8
\end{array}
$$

5.

$$
\begin{array}{r}
{}^{4}\cancel{5}\,{}^{12}3\ \ {}^{1}4 \\
-\ 3\ \ 4\ \ 5 \\
\hline
1\ \ 8\ \ 9
\end{array}
$$

6.

$$
\begin{array}{r}
{}^{4}\cancel{5}\,{}^{12}3\ \ {}^{1}4 \\
-\ 1\ \ 7\ \ 9 \\
\hline
3\ \ 5\ \ 5
\end{array}
$$

7.

$$
\begin{array}{r}
{}^{4}\cancel{5}\,{}^{12}3\ \ {}^{1}4 \\
-\ 2\ \ 6\ \ 8 \\
\hline
2\ \ 6\ \ 6
\end{array}
$$

8.

$$
\begin{array}{r}
{}^{2}\cancel{3}\,{}^{13}\cancel{4}\ \ {}^{1}5 \\
-\ 1\ \ 7\ \ 9 \\
\hline
1\ \ 6\ \ 6
\end{array}
$$

9.

$$
\begin{array}{r}
{}^{2}\cancel{3}\,{}^{13}\cancel{4}\ \ {}^{1}5 \\
-\ 2\ \ 6\ \ 8 \\
\hline
7\ \ 7
\end{array}
$$

10.

$$
\begin{array}{r}
{}^{1}2\,{}^{15}\cancel{6}\ \ {}^{1}8 \\
-\ 1\ \ 7\ \ 9 \\
\hline
8\ \ 9
\end{array}
$$

Check by adding:

1.

$$
\begin{array}{r}
3\ \ 4\ \ 5 \\
+\ 3\ \ 6\ \ 7 \\
\hline
7\ \ 1\ \ 2 \\
{}_{1}\ \ {}_{1}
\end{array}
$$

2.

$$
\begin{array}{r}
1\ \ 7\ \ 9 \\
+\ 5\ \ 3\ \ 3 \\
\hline
7\ \ 1\ \ 2 \\
{}_{1}\ \ {}_{1}
\end{array}
$$

3.

$$
\begin{array}{r}
2\ \ 6\ \ 8 \\
+\ 4\ \ 4\ \ 4 \\
\hline
7\ \ 1\ \ 2 \\
{}_{1}\ \ {}_{1}
\end{array}
$$

4.

$$
\begin{array}{r}
5\ \ 3\ \ 4 \\
+\ 1\ \ 7\ \ 8 \\
\hline
7\ \ 1\ \ 2 \\
{}_{1}\ \ {}_{1}
\end{array}
$$

5.

$$
\begin{array}{r}
3\ \ 4\ \ 5 \\
+\ 1\ \ 8\ \ 9 \\
\hline
5\ \ 3\ \ 4 \\
{}_{1}\ \ {}_{1}
\end{array}
$$

6.

$$
\begin{array}{r}
1\ \ 7\ \ 9 \\
+\ 3\ \ 5\ \ 5 \\
\hline
5\ \ 3\ \ 4 \\
{}_{1}\ \ {}_{1}
\end{array}
$$

7.

$$
\begin{array}{r}
2\ \ 6\ \ 8 \\
+\ 2\ \ 6\ \ 6 \\
\hline
5\ \ 3\ \ 4 \\
{}_{1}\ \ {}_{1}
\end{array}
$$

8.

$$
\begin{array}{r}
1\ \ 7\ \ 9 \\
+\ 1\ \ 6\ \ 6 \\
\hline
3\ \ 4\ \ 5 \\
{}_{1}\ \ {}_{1}
\end{array}
$$

9.

$$
\begin{array}{r}
2\ \ 6\ \ 8 \\
+\quad\ 7\ \ 7 \\
\hline
3\ \ 3\ \ 4 \\
{}_{1}\ \ {}_{1}
\end{array}
$$

10.

$$
\begin{array}{r}
1\ \ 7\ \ 9 \\
+\quad\ 8\ \ 9 \\
\hline
2\ \ 6\ \ 8 \\
{}_{1}\ \ {}_{1}
\end{array}
$$

Rocket 873 − 545 = 328
745 − 417 = 328
545 − 217 = 328

Page 37

Subtracting

1.
```
  5 6  ³ 3 ¹2
      12
-   3 7 5
  ─────────
    2 5 7  m
```

2.
```
  ²3  ⁴ 4 ¹1
      13
-   1 5 4
  ─────────
    1 8 7  m
```

3.
```
  ³4 ¹1 8
-   2 4 2
  ─────────
    1 7 6  m
```

4.
```
    5 2 7
-   3 1 4
  ─────────
    2 1 3  m
```

5.
```
    4 ⁸9 ¹2
-   3 5 6
  ─────────
    1 3 6  m
```

6.
```
    3 9 7
-   1 2 3
  ─────────
    2 7 4  m
```

7. 325 − 187 = 138
8. 362 − 88 = 274
9. 543 − 168 = 375
10. 525 − 377 = 148
11. 613 − 366 = 247
12. 757 − 239 = 518

Rocket Answers will vary.
An example might be: 200 m and 323 m; difference = 123 m

Page 38

Adding and subtracting

1. 328 + 48 = 376 g
2. 454 + 38 = 492 g
3. 562 + 39 = 601 g
4. 434 + 57 = 491 g
5. 669 + 28 = 697 g
6. 375 + 22 = 397 g
7. The answer is always a multiple of 9.
8. 374 − 23 = 351 km
9. 278 − 42 = 236 km
10. 165 − 33 = 132 km
11. 585 − 44 = 541 km
12. 294 − 52 = 242 km
13. 245 − 34 = 211 km

Page 39

Adding and subtracting

1. £2·76 + 18p = £2·94
2. £3·25 + 28p = £3·53
3. £1·75 + 32p = £2·07
4. £2·60 + 15p = £2·75
5. £3·85 + 18p = £4·03
6. £2·80 + 15p = £2·95

Corrections
8. 84 − 53 = 31
10. 254 − 42 = 212
11. 87 − 38 = 49

14. 74 − 35 = 39
16. 65 − 18 = 47
Rocket Answers will vary.

Page 40

Adding and subtracting

1. 138 + 100 = 238 bricks
2. 125 + 15 = 140 bricks
3. 243 + 40 = 283 bricks
4. 176 + 19 = 195 bricks
5. 136 + 169 = 305 bricks
6. 217 + 439 = 656 bricks
7. 87 − 20 = 67
8. 133 − 40 = 93
9. 256 − 120 = 136
10. 84 − 39 = 45
11. 300 − 149 = 151
12. 410 − 101 = 309
Rocket Answers will vary.
13. 47 minutes
14. £260
15. 47 miles
16. 215 miles

Addition and Subtraction PPMs

PPM 49

Pairs to make

Answers will vary

PPM 50

Dice totals

Answers will vary.

PPM 51

All about...

Answers will vary.

PPM 52

Ways to make

Answers will vary.

PPM 53

Take away

Answers will vary.

PPM 54

Take away

Answers will vary.

PPM 55

Count on and back

Answers will vary.

PPM 56

Track to 100

Answers will vary.

PPM 57

Addition pairs

1. 4 + 3 = 7
2. 4 + 4 = 8
3. 6 + 1 = 7
4. 5 + 3 = 8
5. 4 + 1 = 5
6. 6 + 2 = 8
7. 5 + 2 = 7
8. 6 + 3 = 9
9. 5 + 4 = 9
10. 7 + 2 = 9

PPM 58

Adding and subtracting

1. 8 − 3 = 5
2. 3 + 5 = 8
3. 9 − 1 = 8
4. 1 + 8 = 9
5. 8 − 5 = 3
6. 5 + 3 = 8
7. 9 − 4 = 5
8. 4 + 5 = 9
9. 7 − 3 = 4
10. 3 + 4 = 7
11. 8 − 2 = 6
12. 2 + 6 = 8
13. 9 − 6 = 3
14. 6 + 3 = 9

PPM 59

Subtracting

1. 30
2. 40
3. 50
4. 60
5. 69
6. 79
7. 59
8. 70
9. 78
10. 65
11. 59
12. 76

PPM 60

Subtracting in 2s

Sunday: 19
Monday: 17
Tuesday: 15
Wednesday: 13
Thursday: 11
Friday: 9
Saturday: 7

PPM 61

Patterns of 1s and 2s

Total on bus:
18
20
22
23
25
27
28
29
31

PPM 62

Counting in 2s

Children should colour the following numbers:
2, 4, 6, 8, 10, 12, 14, 16, 18, 20.

PPM 63

Patterns of 10s and 5s

Add on in 10s	Add on in 5s	Add on in 10s	Add on in 5s
14	14	7	7
24	19	17	12
34	24	27	17
44	29	37	22
54	34	47	27
64	39	57	32
74	44	67	37
84	49	77	42
94	54	87	47
104	59	97	52

PPM 64

Taking away 10s and 5s

1. Taking away in 10s: 90 80 70 60 50 40 30 20 10 0
2. Taking away in 5s: 90 85 80 75 70 65 60 55 50 45

PPM 65

Counting up a tower

1. First tower: 7 16 25 34 43 52 61 70 79
2. Second tower: 7 17 27 37 47 57 67 77 87
3. Third tower: 7 18 29 40 51 62 73 84 95

PPM 66

Counting up a lighthouse

1. First lighthouse (from bottom): 25 50 75 100 125 150 175 200 225
2. Second lighthouse (from bottom): 25 75 125 175 225 275 325 375 425
3. Third lighthouse (from bottom): 25 125 225 325 425 525 625 725 825

PPM 67

Trio tricks

1. 7 + 3 = 10; 3+ 7 = 10; 10 − 3 = 7; 10 − 7 = 3
   All other answers will vary.

PPM 68

Adding and taking away

1. 8 − 5 = 3
2. 6 − 4 = 2
3. 7 − 2 = 5
4. 9 − 5 = 4
5. 9 − 3 = 6

6.–8. Answers will vary.

PPM 69

Adding machines

1.

In	23	29	38	17	46	19
out	27	33	42	21	50	23

2.

In	37	26	18	45	59	46
out	52	41	33	60	74	41

3.

In	13	37	24	48	66	79
out	20	44	31	55	73	86

PPM 70

Subtracting 10s

1. 17 − 10 = 7
2. 28 − 10 = 18
3. 35 − 30 = 5
4. 44 − 20 = 24
5. 21 − 10 = 11
6. 53 − 40 = 13
7. 42 − 30 = 12
8. 39 − 20 = 19
9. 60 − 50 = 10
10. 56 − 20 = 36

PPM 71

Doubles

1. 2 + 2 = 4
2. 1 + 1 = 2
   Children should draw 1 block.
3. 3 + 3 = 6
   Children should draw a tower of 3 blocks.
4. 4 + 4 = 8
   Children should draw a tower of 4 blocks.
5. 5 + 5 = 10
   Children should draw a tower of 5 blocks.
6. 6 + 6 = 12
   Children should draw a tower of 6 blocks.
7. 7 + 7 = 14
   Children should draw a tower of 7 blocks.
8. 8 + 8 = 16
   Children should draw a tower of 8 blocks.

PPM 72

Doubles

Children should colour the items in questions 1, 4, 5, 6, 7, 9, 10.

1. 5 + 5 = 10
2. 3 + 4 = 7
3. 2 + 1 = 3
4. 3 + 3 = 6
5. 10 + 10 = 20
6. 5 + 5 = 10
7. 4 + 4 = 8
8. 3 + 2 = 5
9. 6 + 6 = 12
10. 7 + 7 = 14

PPM 73

Doubling

1.

In	9	10	8	5	1	3	6	4	7	2
out	18	20	16	10	2	6	12	8	14	4

2.

In	10	25	40	5	50	30	45	15	35	20
out	20	50	80	10	100	60	90	30	70	40

PPM 74

Doubling

1. 5
2. 9
3. 7
4. 11
5. 17
6. 15

PPM 75

Making the next 10

1. 69 + 1 = 70
2. 37 + 3 = 40
3. 26 + 4 = 40
4. 52 + 8 = 60
5. 81 + 9 = 90
6. 45 + 5 = 50
7. 74 + 6 = 80
8. 13 + 7 = 20
9. 32 + 8 = 40
10. 88 + 2 = 90

PPM 76

Adding to 100

1. 60 + 40 = 100
2. 80 + 20 = 100
3. 70 + 30 = 100
4. 90 + 10 = 100

5. $55 + 45 = 100$
6. $40 + 60 = 100$
7. $10 + 90 = 100$
8. $30 + 70 = 100$
9. $75 + 25 = 100$
10. $65 + 35 = 100$
11. $5 + 95 = 100$
12. $25 + 75 = 100$

PPM 77
Using a 100 square

	Make 40	Make 50	Make 60
38	2	12	22
31	9	19	29
25	15	25	35
19	21	31	41
5	35	45	55
16	24	34	44
7	33	43	53
34	6	16	26
28	12	22	32
35	5	15	25
27	13	23	33
39	1	11	21
15	25	35	45
26	14	24	34
4	36	46	56
23	17	27	37

PPM 78
Subtracting

1. $24 - 4 = 20$
$24 - 6 = 18$
2. $43 - 3 = 40$
$43 - 5 = 38$
3. $23 - 3 = 20$
$23 - 6 = 17$
4. $33 - 3 = 30$
$33 - 7 = 26$
5. $41 - 1 = 40$
$41 - 4 = 37$
6. $54 - 4 = 50$
$54 - 7 = 47$
7. $62 - 2 = 60$
$62 - 5 = 57$

PPM 79
Differences

1. 3
2. 3
3. 4
4. 3
5. 8
6. 3
7. 4
8. 6
9. 6

PPM 80
Partitioning using grids

1. $50 + 3$
2. $40 + 2$
3. $20 + 5$
4. $20 + 6$
5. $40 + 3$
6. $80 + 2$
7. $20 + 3$
8. $10 + 5$
9. $30 + 6$
10. $60 + 3$
11. $50 + 40 + 3 + 2 = 90 + 5 = 95$
12. $20 + 20 + 5 + 3 = 40 + 8 = 48$
13. $20 + 10 + 3 + 5 = 30 + 8 = 38$
14. $20 + 40 + 6 + 3 = 60 + 9 = 69$

PPM 81
Partitioning using grids

1. $200 + 50 + 3 + 300 + 40 + 2$
$= 500 + 90 + 5$
$= 595$
2. $300 + 10 + 5 + 200 + 70 + 3$
$= 500 + 80 + 8$
$= 588$
3. $400 + 20 + 6 + 300 + 70 + 1$
$= 700 + 90 + 7$
$= 797$
4. $200 + 30 + 5 + 400 + 10 + 2$
$= 600 + 40 + 7$
$= 647$
5. $500 + 60 + 3 + 200 + 6$
$= 700 + 60 + 9$
$= 769$
6. $\underline{123} + \underline{246}$
$= 100 + 20 + 3 + 200 + 40 + 6$
$= 369$

PPM 82
Adding several numbers

1.

3	1	6	10
5	7	2	14
8	4	3	15
16	**12**	**11**	

2.

8	5	3	16
2	7	2	11
1	4	6	11
11	**16**	**11**	

3.

3	7	1	11
8	2	6	16
3	5	4	12
14	**14**	**11**	

4.

3	6	2	11
8	1	5	14
9	7	4	20
20	**14**	**11**	

PPM 83
Adding several numbers

1. $4 + 6 + 5 = 15$
2. $7 + 3 + 8 = 18$
3. $6 + 4 + 6 = 16$
4. $3 + 7 + 9 = 19$
5. $5 + 8 = 13$ $13 + 9 = 22$
6. $4 + 7 = 11$ $11 + 9 = 20$
7. $6 + 8 = 14$ $14 + 9 = 23$
8. $7 + 7 = 14$ $14 + 9 = 23$
9. $8 + 8 = 16$ $16 + 9 = 25$
10. $4 + 6 + 3 = 13$ $13 + 9 = 22$
11. $7 + 3 + 4 = 14$ $14 + 9 = 23$
12. $13 + 8 = 21$ $21 + 9 = 30$

PPM 84
Taking away

1. $34 - 8 - 7 = 19$
2. $62p - 5p - 4p - 12p = 41p$
3. Answers will vary.

PPM 85
Adding

1. $4 + 7 = 11$ $14 + 7 = 21$
$44 + 7 = 51$
2. $6 + 6 = 12$ $26 + 6 = 32$
$56 + 6 = 62$
3. $9 + 5 = 14$ $39 + 5 = 44$
$79 + 5 = 84$
4. $4 + 8 = 12$ $54 + 8 = 62$
$84 + 8 = 92$
5. $7 + 8 = 15$ $7 + 38 = 45$
$7 + 68 = 75$
6. $5 + 13 = 18$ $5 + 43 = 48$
$5 + 73 = 78$
7. $18 - 9 = 9$ $48 - 9 = 39$
$78 - 9 = 69$
8. $14 - 6 = 8$ $74 - 6 = 68$
$94 - 6 = 88$
9. $15 - 7 = 8$ $55 - 7 = 48$
$85 - 7 = 78$
10. $13 - 7 = 6$ $73 - 7 = 66$
$53 - 7 = 46$
11–14. Answers will vary.

PPM 86
Subtracting close numbers

1. 70p
2. 79p
3. 57p
4. 59p
5. 68p
6. 88p

7.

Number pair		How much bigger?	Total
63	4	59	67
28	5	23	33
7	52	45	59
81	8	73	89
23	7	16	30
9	48	39	57
4	71	67	75
98	9	89	107

PPM 87

Number triangles

1.

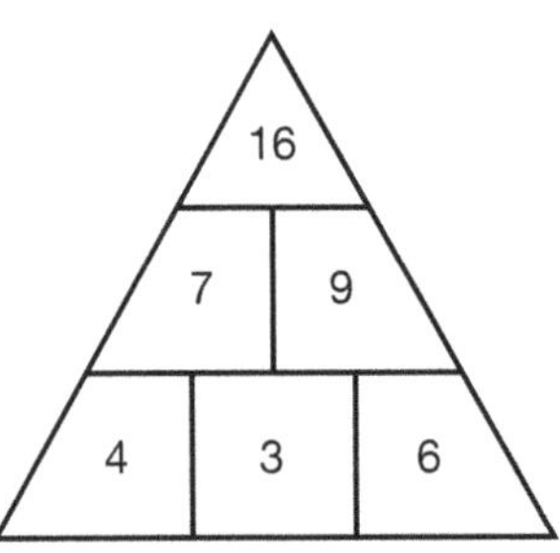

2.

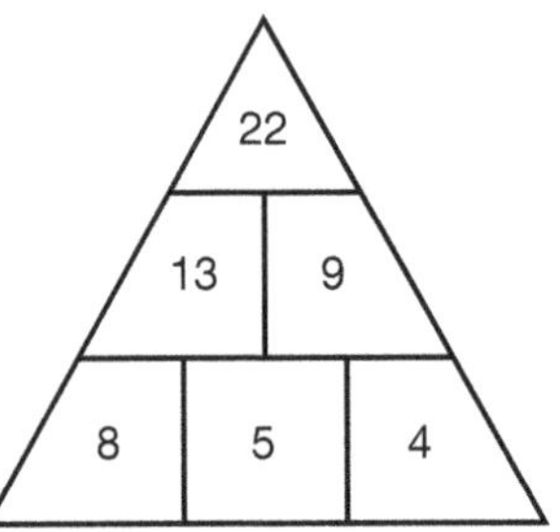

3.

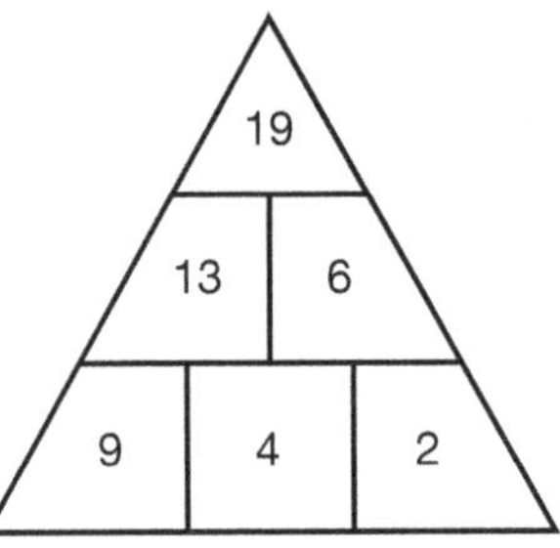

4.

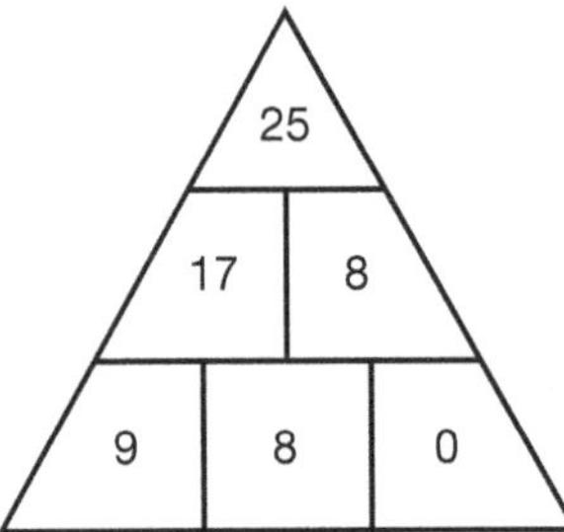

5.

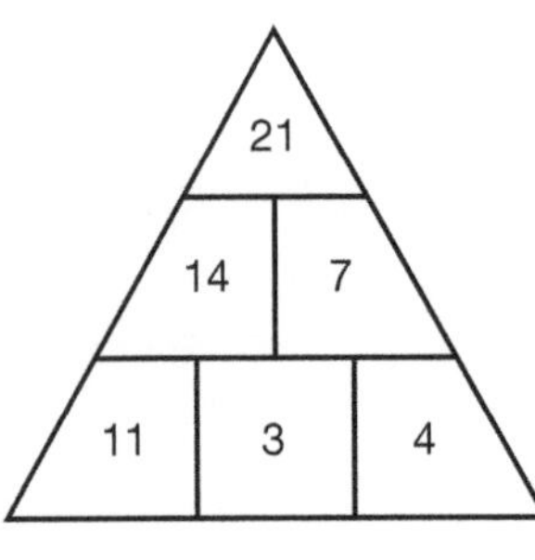

6.

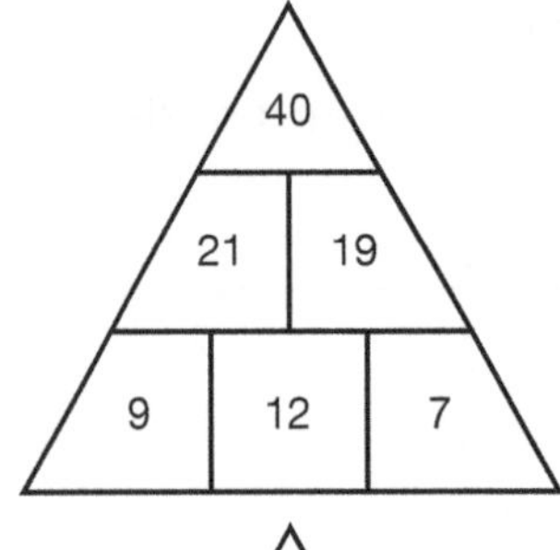

7.

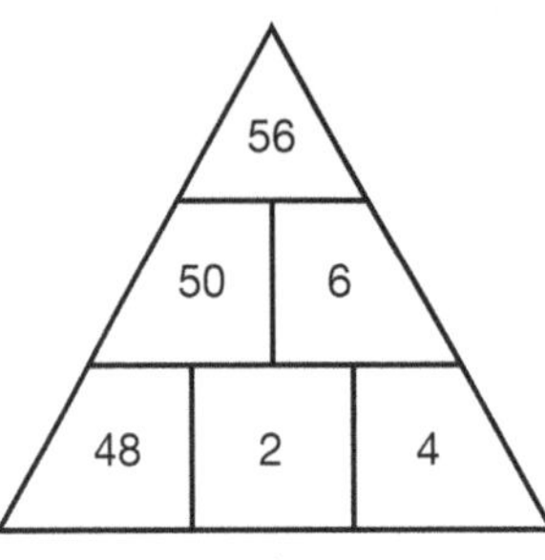

8.

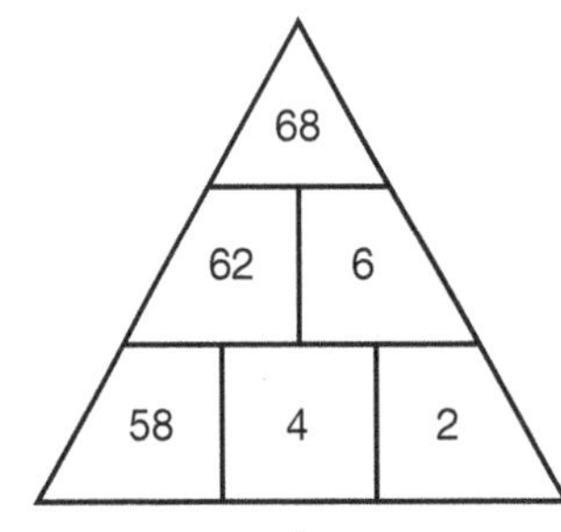

9.

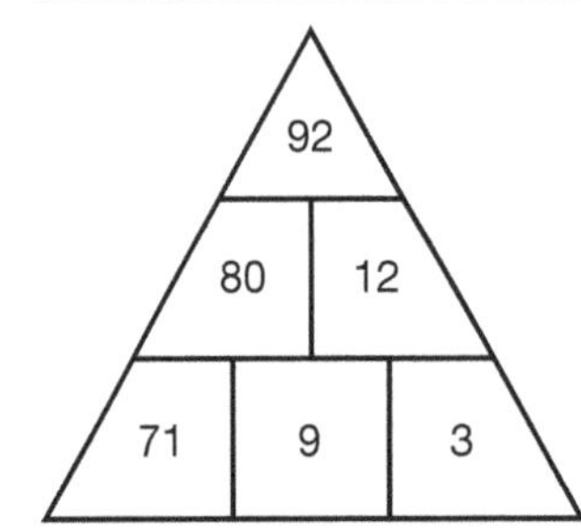

PPM 88

Subtracting (calculating)

1. $36 + 24 = 60$
2. $28 + 22 = 50$
3. $46 + 24 = 70$
4. $65 + 35 = 90$
5. $37 + 24 = 61$
6. $28 + 45 = 73$
7. $52 + 34 = 86$
8. $23 + 35 = 58$

PPM 89

Adding and subtracting

1. $224 + 29 = 253$
2. $342 + 19 = 361$
3. $148 + 29 = 177$
4. $563 + 21 = 584$
5. $236 + 19 = 255$
6. $472 + 21 = 493$
7. £346 − £19 = £327
8. £542 − £19 = £523
9. £638 − £19 = £619
10. £456 − £21 = £435
11. £164 − £21 = £143
12. £288 − £31 = £257

PPM 90

Adding

1. $17p + 10p = 27p$
2. $25p + 30p = 55p$
3. $41p + 40p = 81p$
4. $54p + 30p = 84p$
5. $74p + 20p = 94p$
6. $58p + 40p = 94p$
7. $60p + 30p = 90p$
8. $32p + 30p = 62p$
9. $79p + 20p = 99p$
10. $50p + 30p = 80p$
11. $16p + 50p = 66p$
12. $57p + 10p = 67p$
13. $23p + 40p = 63p$

PPM 91

Adding

1. $10 + 82 = 92$
2. $64 + 60 = 124$
3. $57 + 40 = 97$
4. $66 + 30 = 96$
5. $78 + 70 = 148$
6. $54 + 40 = 94$
7. $46 + 70 = 116$
8. $54 + 60 = 114$
9. $82 + 50 = 132$
10. $146 + 30 = 176$
11. $173 + 20 = 193$
12. $155 + 40 = 195$
13. $261 + 30 = 291$
14. $342 + 50 = 392$

PPM 92

Adding and subtracting

1. $274 + 40 = 314\,m$
2. $364 + 50 = 414\,m$
3. $482 + 20 = 502\,m$
4. $275 + 50 = 325\,m$
5. $355 + 40 = 395\,m$
6. $568 + 30 = 598\,m$
7. $284 − 40 = 244\,m$
8. $361 − 30 = 331\,m$
9. $568 − 50 = 518\,m$
10. $486 − 60 = 426\,m$
11. $377 − 40 = 337\,m$
12. $295 − 50 = 245\,m$

PPM 93

Formal addition methods

1.
$$\begin{array}{r} 9\ 6 \\ +\ \underline{1\ 8} \\ 1\ 1\ 4 \\ {}^{1} \end{array}$$

2.
$$\begin{array}{r} 5\ 6 \\ +\ \underline{8\ 8} \\ 1\ 4\ 4 \\ {}^{1} \end{array}$$

3.
$$\begin{array}{r} 2\ 8 \\ +\ \underline{9\ 5} \\ 1\ 2\ 3 \\ {}^{1} \end{array}$$

4.
```
    9 2
 +  4 9
  1 4 1
    1
```

5.
```
    9 7
 +  7 5
  1 7 2
    1
```

6.
```
    5 3
 +  5 8
  1 1 1
    1
```

7.
```
    7 8
 +  7 3
  1 5 1
    1
```

8.
```
    3 9
 +  2 4
    6 3
    1
```

9.
```
    9 4
 +  5 7
  1 5 1
    1
```

10.
```
    7 8
 +  2 4
  1 0 2
    1
```

11.
```
    7 9
 +  3 8
  1 1 7
    1
```

12.
```
    8 5
 +  7 7
  1 6 2
    1
```

PPM 94
Adding three numbers

1.
```
    4 6
    2 4
 +  3 3
  1 0 3
    1
```

2.
```
    3 7
    3 3
 +  4 2
  1 1 2
    1
```

3.
```
    2 8
    5 1
 +  1 9
    9 8
    1
```

4.
```
    6 3
    4 7
 +  5 5
  1 6 5
    1
```

5.
```
    4 9
    8 1
 +  2 9
  1 5 9
    1
```

6.
```
    5 5
    6 5
 +  3 1
  1 5 1
    1
```

7.
```
    2 7
    8 5
 +  4 3
  1 5 5
    1
```

8.
```
    4 8
    6 3
 +  2 9
  1 4 0
    2
```

9.
```
    1 9
    8 1
 +  7 7
  1 7 7
    1
```

10.
```
    4 8
    5 4
 +  7 6
  1 7 8
    1
```

11.
```
    8 4
    2 7
 +  9 3
  2 0 4
    1
```

12.
```
    5 2
    6 9
 +  3 5
  1 5 6
    1
```

PPM 95
Making totals

1.
```
    2 5           2 5
    5 3           4 6
 +  2 7    or  +  3 4
  1 0 5         1 0 5
    1             1
```

2.
```
    3 1           3 1
    5 3           4 6
 +  2 7    or  +  3 4
  1 1 1         1 1 1
    1             1
```

3.
```
    2 5
    5 3
 +  3 4
  1 1 2
    1
```

4.
```
    2 5
    2 7
 +  3 4
    8 6
    1
```

5.
```
    4 2
    5 3
 +  4 6
  1 4 1
    1
```

6.
```
    3 1
    2 7
 +  3 4
    9 2
    1
```

PPM 96
Adding using formal methods

1.
```
  H T U
  1 3 3
  2 4 2
+ 3 2 6
  7 0 1
  1 1
```

2.
```
  H T U
  3 1 1
  1 3 9
+ 2 3 4
  6 8 4
    1
```

3.
```
  H T U
  3 6 2
  3 2 6
+ 1 1 3
  8 0 1
  1 1
```

4.
```
  H T U
  3 4 6
  2 4 1
+ 1 1 5
  7 0 2
  1 1
```

5.
```
  H T U
  3 1 0
  2 4 9
+ 1 3 3
  6 9 2
    1
```

6.
```
  H T U
  2 2 2
  3 6 4
+ 1 1 9
  7 0 5
  1 1
```

7.
```
  H T U
  1 2 7
  1 5 8
+ 4 1 2
  6 9 7
    1
```

8.
```
  H T U
  4 4 4
  2 2 1
+ 1 3 7
  8 0 2
  1 1
```

9.
```
  H T U
  2 4 4
  1 2 2
+ 4 1 8
  7 8 4
    1
```

10.
```
  H T U
  4 2 3
  1 4 9
+ 2 2 6
  7 9 8
    1
```

11.
```
  H T U
  4 1 6
  2 2 7
+ 1 6 6
  8 0 9
  1 1
```

12.

```
  H T U
  1 5 4
  4 1 5
+ 2 3 9
───────
  8 0 8
  1 1
```

PPM 97
Adding using formal methods
Rounds 1–6 Answers will vary.

PPM 98
Subtracting
Children should include estimates.

1.
```
  8 4
− 2 3
─────
  6 1
```

2.
```
  ⁵6̶ ¹2
−   3 6
───────
    2 6
```

3.
```
  ⁶7̶ ¹5
−   1 6
───────
    5 9
```

4.
```
  ⁸9̶ ¹3
−   3 7
───────
    5 6
```

5.
```
  ³4̶ ¹5
−   2 8
───────
    1 7
```

6.
```
  ⁷8̶ ¹2
−   5 7
───────
    2 5
```

7.
```
  ⁶7̶ ¹4
−   5 6
───────
    1 8
```

8.
```
  ⁵6̶ ¹1
−   2 7
───────
    3 4
```

9.
```
  ⁷8̶ ¹7
−   2 8
───────
    5 9
```

PPM 99
Row and column totals

1.
```
  9 2
− 1 6
─────
    7
```

2.
```
  9 2
− 2 4
─────
    6
```

3.
```
  9 2
− 2 0
─────
    7
```

4.
```
  9 2
− 2 7
─────
    6
```

5.
```
  9 2
− 1 8
─────
    7
```

6.
```
  9 2
− 2 3
─────
    6
```

7.
```
  9 2
− 2 4
─────
    6
```

8.
```
  9 2
− 2 2
─────
    7
```

PPM 100
Problems

1. 168. 32.
2. 229. 183.
3. 219 days. 31 days.
4. 116.
5. 325. 650.
6. 53p. 52p.
7. 147. 3.
8. £76.

PPM 101
Problems

1. 6 years
2. 8
3. 7p
4. 121 cm
5. 19
6. 22
7. Answers will vary.

Addition and Subtraction APMs

APM 208
Number pyramids

1.
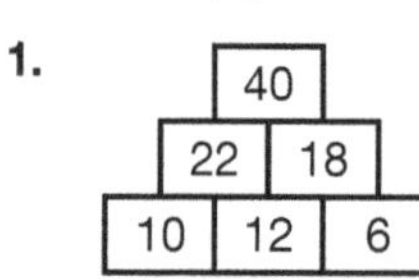

2.
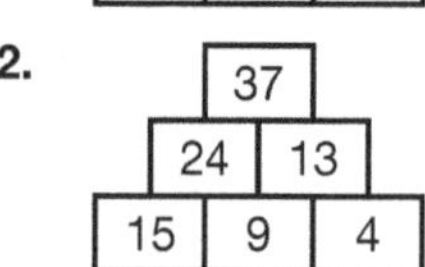

3.
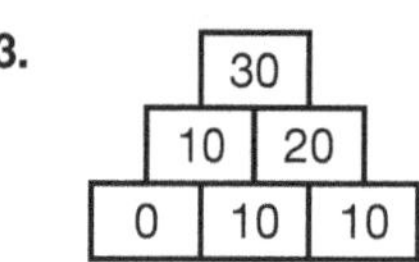

4.
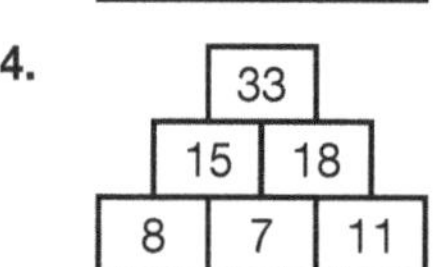

5.
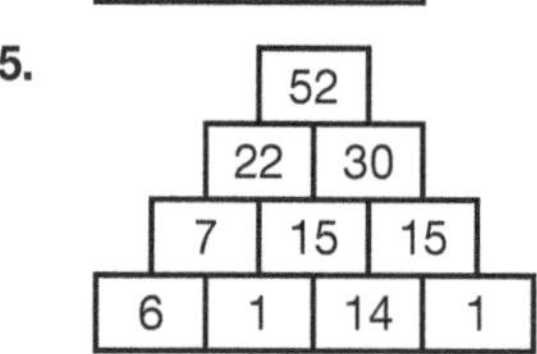

6. Answers will vary

APM 210
More number pyramids

1.
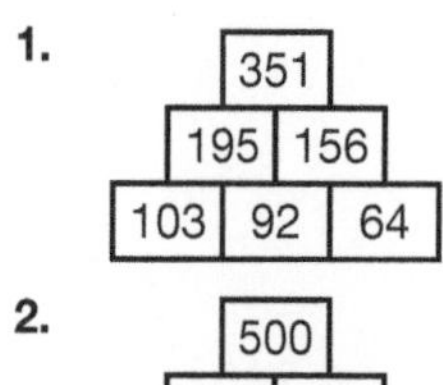

2.

3.

4.

5.

6. Answers will vary

APM 211
Easy or hard?

345 − 123 = 222
567 − 475 = 92
204 − 97 = 107
1689 − 345 = 1344
666 − 333 = 333
4678 − 2346 = 2332
5764 − 1379 = 4385
4005 − 1984 = 2021
7350 − 6999 = 351
2345 − 1879 = 466
56 784 520 895 643 − 12 543 110 562 412
= 44 241 410 333 231

Multiplication and Division Pupil Book 3

Page 3
Multiplying

1. $3 \times 3 = 9$
2. $2 \times 4 = 8$
3. $5 \times 2 = 10$
4. $3 \times 5 = 15$
5. $4 \times 4 = 16$
6. $3 \times 6 = 18$
7. 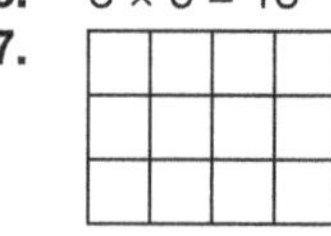
$3 \times 4 = 12$
8. 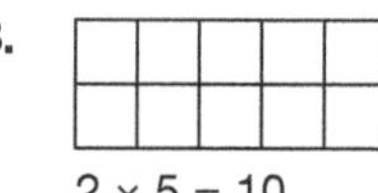
$2 \times 5 = 10$
9. 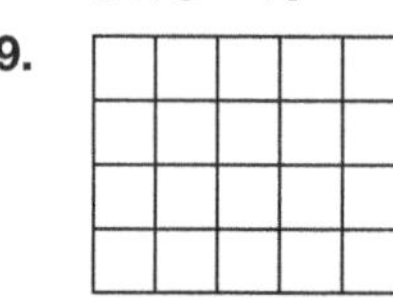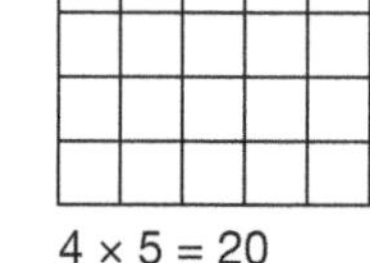
$4 \times 5 = 20$
10. 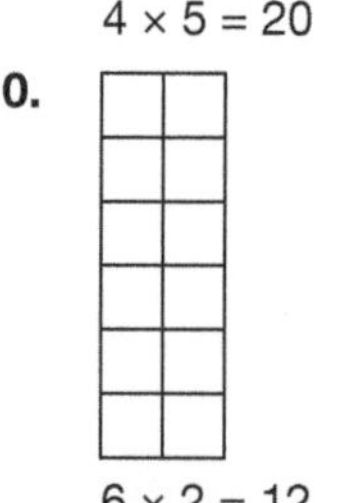
$6 \times 2 = 12$

11. 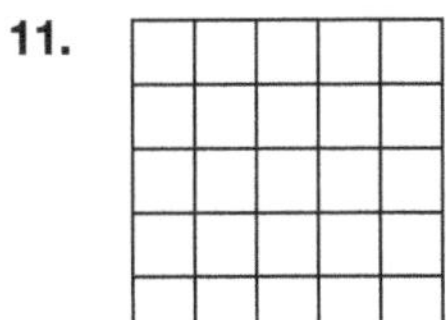
5 × 5 = 25

12. 
3 × 10 = 30

13. 
2 × 2 = 4

14. 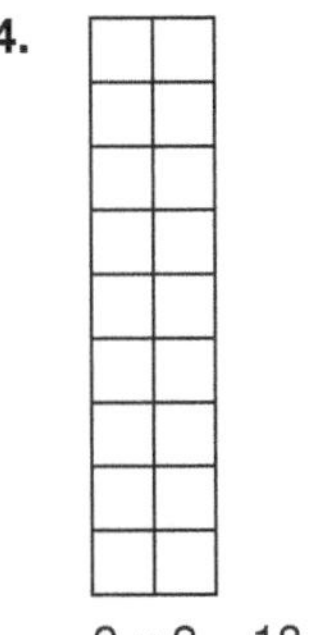
9 × 2 = 18

Rocket Answers will vary.

Page 4

Multiplying

1. 3 × 5 = 15
2. 2 × 4 = 8
3. 3 × 3 = 9
4. 2 × 6 = 12
5. 4 × 5 = 20
6. 2 × 7 = 14
7. 3 × 4 = 12
8. 3 × 6 = 18
9. 5 × 5 = 25
10. 3 × 4 = 12

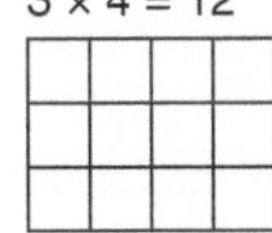
4 × 3 = 12

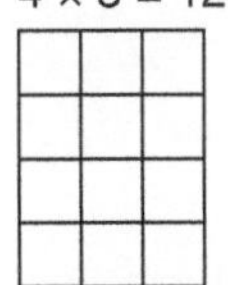

11. 4 × 5 = 20

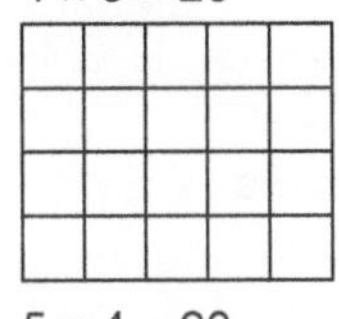
5 × 4 = 20

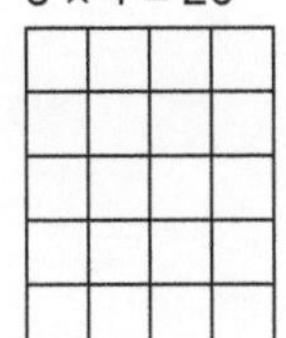

12. 3 × 5 = 15

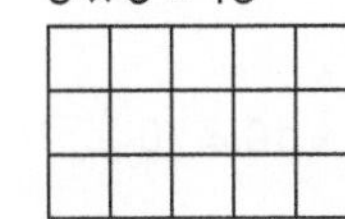

---

5 × 3 = 15
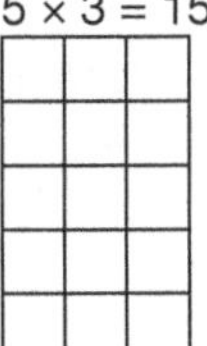

13. 2 × 3 = 6
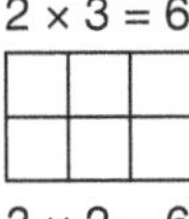

3 × 2 = 6
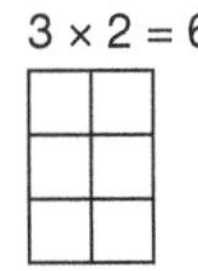

Rocket

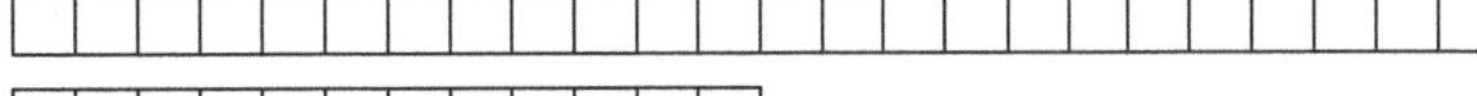

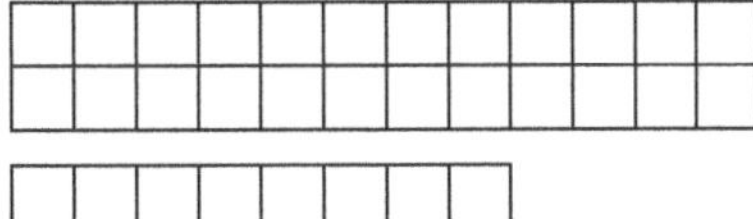

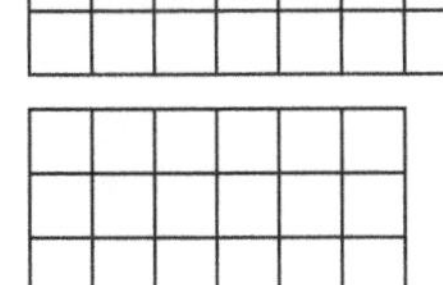

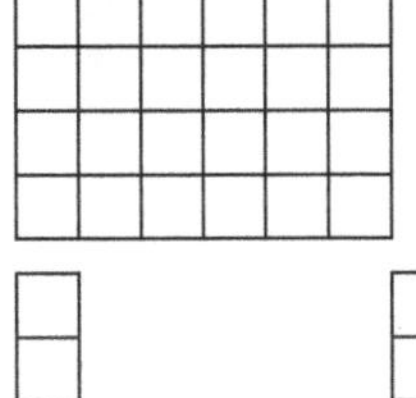

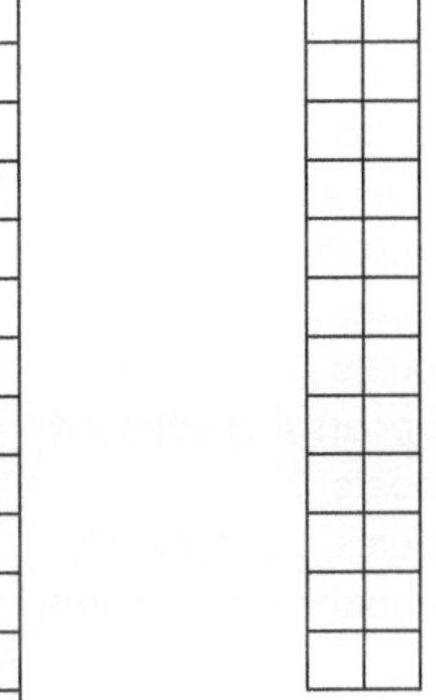

Page 5

Multiplying

1. 1 and 4, 2 and 8, 3 and 6, 5 and 7, 9 and 10

Rocket Children should draw rectangles representing: 1 × 12, 12 × 1, 2 × 6, 6 × 2, 3 × 4, 4 × 3.

11. 2 × 4 = 8
12. 5 × 3 = 15
13. 2 × 8 = 16
14. 6 × 4 = 24
15. 3 × 7 = 21
16. 9 × 2 = 18
17. 3 × 6 = 18
18. 4 × 5 = 20
19. 7 × 6 = 42

Page 6

Multiplying

1. 2 × 4 = 8, 4 × 2 = 8
2. 4 × 3 = 12, 3 × 4 = 12
3. 3 × 5 = 15, 5 × 3 = 15
4. 6 × 2 = 12, 2 × 6 = 12
5. 2 × 5 = 10, 5 × 2 = 10
6. 4 × 6 = 24, 6 × 4 = 24
7. 6 × 3 = 18, 3 × 6 = 18
8. 5 × 4 = 20, 4 × 5 = 20
9. 3 × 3 = 9
10. 4 × 2 = 8
11. 5 × 5 = 25
12. 5 × 2 = 10
13. 3 × 5 = 15
14. 4 × 3 = 12

Rocket Children should draw rectangles representing: 3 × 4, 4 × 3, 4 × 4, 2 × 5, 5 × 2, 3 × 5, 5 × 3, 4 × 5, 5 × 4, 2 × 6, 6 × 2, 3 × 6, 6 × 3.

Page 7

Multiplying and dividing

1. 6 ÷ 3 = 2
2. 10 ÷ 2 = 5 or 10 ÷ 5 = 2
3. 12 ÷ 4 = 3 or 12 ÷ 3 = 4
4. 20 ÷ 5 = 4 or 20 ÷ 4 = 5

Multiplications for each set

1. 2 × 3 = 6 or 3 × 2 = 6
2. 5 × 2 = 10 or 2 × 5 = 10
3. 3 × 4 = 12 or 4 × 3 = 12
4. 4 × 5 = 20 or 5 × 4 = 20

Children should draw objects set in the following ways:

5. 14 objects into 2 groups of 7
6. 9 objects into 3 groups of 3
7. 15 objects into 5 groups of 3
8. 20 objects into 4 groups of 5
9. 10 objects into 2 groups of 5
10. 24 objects into 6 groups of 4
11. 30 objects into 10 groups of 3
12. 25 objects into 5 groups of 5.
13. 4 × 3 = 12, 12 ÷ 3 = 4
14. 2 × 4 = 8, 8 ÷ 4 = 2
15. 3 × 5 = 15, 15 ÷ 5 = 3
16. 5 × 2 = 10, 10 ÷ 2 = 5

Page 8

Multiplying by 2

1. 12
2. 8
3. 2
4. 14
5. 6
6. 18

Rocket Answers will vary, however all positions marked should be multiples of 20.

7. $4 \times 2 = 8$
8. $7 \times 2 = 14$
9. $5 \times 2 = 10$
10. $3 \times 2 = 6$
11. $8 \times 2 = 16$
12. $9 \times 2 = 18$
13. $5 \times 2 = 10$
14. $10 \times 2 = 20$
15. $6 \times 2 = 12$
16. $3 \times 2 = 6$
17. $2 \times 2 = 4$
18. $9 \times 2 = 18$

Page 9

Fours

1	2	3	4
5	6	7	8
9	10	11	12
13	14	15	16
17	18	19	20
21	22	23	24
25	26	27	28
29	30	31	32
33	34	35	36
37	38	39	40

1. 8
2. 40
3. 20
4. 16
5. 12
6. 28
7. 24
8. 36
9. $3 \times 4 = 12$
10. $5 \times 4 = 20$
11. $2 \times 4 = 8$
12. $4 \times 4 = 16$
13. $10 \times 4 = 40$
14. $7 \times 4 = 28$

Rocket Answers will vary.

Page 10

Fours

1. $3 \times 2 = 6, 3 \times 4 = 12$
2. $5 \times 2 = 10, 5 \times 4 = 20$
3. $7 \times 2 = 14, 7 \times 4 = 28$
4. $4 \times 2 = 8, 4 \times 4 = 16$
5. $9 \times 2 = 18, 9 \times 4 = 36$
6. $6 \times 2 = 12, 6 \times 4 = 24$

7. $8 \times 2 = 16, 8 \times 4 = 32$

Rocket Repeat the pattern used above, for example, $1 \times 4 = 4$, $1 \times 8 = 8$; $2 \times 4 = 8, 2 \times 8 = 16$, and so on.

8. $5 \times 4 = 20$
9. $6 \times 4 = 24$
10. $2 \times 4 = 8$
11. $3 \times 4 = 12$
12. $10 \times 4 = 40$
13. $7 \times 4 = 28$
14. $4 \times 4 = 16$
15. $9 \times 4 = 36$
16. $6 \times 4 = 24$
17. $5 \times 4 = 20$
18. $8 \times 4 = 32$

Page 11

Eights

1. $4 \times 8 = 32$
2. $3 \times 8 = 24$
3. $5 \times 8 = 40$
4. $6 \times 8 = 48$
5. $8 \times 8 = 64$
6. $7 \times 8 = 56$
7. $16 \div 8 = 2$ boats
8. $48 \div 8 = 6$ boats
9. $56 \div 8 = 7$ boats
10. $24 \div 8 = 3$ boats
11. $32 \div 8 = 4$ boats
12. $80 \div 8 = 10$ boats

Rocket
7. 3 boats 2 spaces left
8. 8 boats
9. 10 boats 4 spaces left
10. 4 boats
11. 6 boats 4 spaces left
12. 14 boats 4 spaces left

Page 12

Tens

1. $4 \times 10 = 40$
2. $3 \times 10 = 30$
3. $5 \times 10 = 50$
4. $6 \times 10 = 60$
5. $8 \times 10 = 80$
6. $7 \times 10 = 70$
7. 2 buses
8. 5 buses
9. 8 buses
10. 3 buses
11. 7 buses
12. 10 buses

Rocket
7. 3 buses 4 spaces left
8. 7 buses 6 spaces left
9. 10 buses
10. 4 buses 2 spaces
11. 9 buses 2 spaces
12. 13 buses 4 spaces

Page 13

Fives and tens

1. $3 \times 10 = 30$

2. $5 \times 10 = 50$
3. $4 \times 10 = 40$
4. $7 \times 10 = 70$
5. $10 \times 10 = 100$
6. $6 \times 10 = 60$
7. $9 \times 10 = 90$

Rocket Answers will vary.

8. $5 \times 5 = 25$
9. $2 \times 5 = 10$
10. $6 \times 5 = 30$
11. $3 \times 5 = 15$
12. $10 \times 5 = 50$
13. $7 \times 5 = 35$
14. $8 \times 5 = 40$
15. $4 \times 5 = 20$
16. $9 \times 5 = 45$

Page 14

Fives and tens

1. $3 \times 5p = 15p$
2. $5 \times 10p = 50p$
3. $5 \times 5p = 25p$
4. $7 \times 10p = 70p$
5. $4 \times 10p = 40p$
6. $6 \times 5p = 30p$
7. $8 \times 5p = 40p$
8. $6 \times 10p = 60p$

Rocket

1. 85p
2. 50p
3. 75p
4. 30p
5. 60p
6. 70p
7. 60p
8. 40p
9. 15
10. 30
11. 10
12. 45
13. 35
14. 20
15. 25
16. 40

Page 15

Fives and tens

1. $4 \times 5p = 20p$
2. $2 \times 5p = 10p$
3. $7 \times 5p = 35p$
4. $10 \times 5p = 50p$
5. $20 \times 5p = 100p$ or £1
6. $6 \times 5p = 30p$

Rocket 1p = 0 throws
2p = 0 throws
5p = 1 throw
10p = 2 throws
50p = 10 throws
£1 = 20 throws
£2 = 40 throws

7. $4 \times 5 = 20$
8. $6 \times 10 = 60$
9. $7 \times 5 = 35$
10. $4 \times 10 = 40$
11. $6 \times 5 = 30$
12. $8 \times 5 = 40$
13. $7 \times 10 = 70$
14. $9 \times 10 = 90$
15. $7 \times 5 = 35 \rightarrow 7 \times 10 = 70$
half of 70 = 35

16. $12 \times 5 = 60 \rightarrow 12 \times 10 = 120$
half of $120 = 60$
17. $21 \times 5 = 105 \rightarrow 21 \times 10 = 210$
half of $210 = 105$
18. $32 \times 5 = 160 \rightarrow 32 \times 10 = 320$
half of $320 = 160$
19. $46 \times 5 = 230 \rightarrow 46 \times 10 = 460$
half of $460 = 230$
20. $14 \times 5 = 70 \rightarrow 14 \times 10 = 140$
half of $140 = 70$

Page 16

Threes

1. $4 \times 3 = 12$
2. $6 \times 3 = 18$
3. $3 \times 3 = 9$
4. $10 \times 3 = 30$
5. $8 \times 3 = 24$
6. $5 \times 3 = 15$
7. $9 \times 3 = 27$
8. $2 \times 3 = 6$
9. $7 \times 3 = 21$
10. $9 \div 3 = 3$
11. $15 \div 3 = 5$
12. $27 \div 3 = 9$
13. $6 \div 3 = 2$
14. $30 \div 3 = 10$
15. $21 \div 3 = 7$
16. $12 \div 3 = 4$
17. $24 \div 3 = 8$
18. $18 \div 3 = 6$

Rocket 180

Page 17

Threes and sixes

1. $5 \times 3 = 15$
2. $6 \times 3 = 18$
3. $2 \times 3 = 6$
4. $4 \times 3 = 12$
5. $9 \times 3 = 27$
6. $3 \times 3 = 9$

1	2	3	4	5	6
7	8	9	10	11	12
13	14	15	16	17	18
19	20	21	22	23	24
25	26	27	28	29	30
31	32	33	34	35	36
37	38	39	40	41	42
43	44	45	46	47	48
49	50	51	52	53	54
55	56	57	58	59	60

7. 24
8. 42
9. 12
10. 60
11. 30
12. 36

Rocket 66, 72, 78, 84, 90, 96, 102, 108, 114, 120

Page 18

Sixes

1. $3 \times 6 = 18$
2. $5 \times 6 = 30$
3. $2 \times 6 = 12$
4. $7 \times 6 = 42$
5. $10 \times 6 = 60$
6. $8 \times 6 = 48$
7. $6 \times 6 = 36$
8. $9 \times 6 = 54$
9. $4 \times 6 = 24$

Rocket 42 and 48

10. $18 \div 6 = 3$ boxes
11. $30 \div 6 = 5$ boxes
12. $54 \div 6 = 9$ boxes
13. $36 \div 6 = 6$ boxes
14. $24 \div 6 = 4$ boxes
15. $42 \div 6 = 7$ boxes

Page 19

Sixes

1. 24
2. 36
3. 18
4. 54
5. 12
6. 30
7. 42
8. 48
9. 7 bags 5 oranges left over
10. 58p
11. 54 days

Rocket Answers will vary

12. $36 \div 6 = 6$
13. $18 \div 6 = 3$
14. $42 \div 6 = 7$
15. $12 \div 6 = 2$
16. $54 \div 6 = 9$
17. $60 \div 6 = 10$
18. $24 \div 6 = 4$
19. $30 \div 6 = 5$
20. $48 \div 6 = 8$

Page 20

Sixes

1.

3s	3	6	9	12	15	18	21	24	27	30
6s	6	12	18	24	30	36	42	48	54	60

2.

1s	1	2	3	4	5	6	7	8	9	10
5s	5	10	15	20	25	30	35	40	45	50
6s	6	12	18	24	30	36	42	48	54	60

3.

2s	2	4	6	8	10	12	14	16	18	20
4s	4	8	12	16	20	24	28	32	36	40
6s	6	12	18	24	30	36	42	48	54	60

Rocket

3s	3	6	9	12	15	18	21	24	27	30
6s	6	12	18	24	30	36	42	48	54	60
9s	9	18	27	36	45	54	63	72	81	90

To find the 9s children could also multiply the 3s by 3, or add 4s and 5s

4. $4 \times 3 = 12 \rightarrow 4 \times 6 = 24$
5. $7 \times 3 = 21 \rightarrow 7 \times 6 = 42$
6. $9 \times 3 = 27 \rightarrow 9 \times 6 = 54$
7. $11 \times 3 = 33 \rightarrow 11 \times 6 = 66$
8. $3 \times 6 = 18$
9. $8 \times 6 = 48$
10. $2 \times 6 = 12$
11. $5 \times 6 = 30$
12. $6 \times 6 = 36$
13. $20 \times 6 = 120$

Page 21

Multiplying and dividing

1. $12 \div 3 = 4$ or $12 \div 4 = 3$
2. $8 \div 4 = 2$ or $8 \div 2 = 4$
3. $15 \div 5 = 3$ or $15 \div 3 = 5$
4. $20 \div 4 = 5$ or $20 \div 5 = 4$
5. $18 \div 3 = 6$ or $18 \div 6 = 3$
6. $16 \div 4 = 4$

Multiplications:

1. $3 \times 4 = 12, 4 \times 3 = 12$
2. $4 \times 2 = 8, 2 \times 4 = 8$
3. $5 \times 3 = 15, 3 \times 5 = 15$
4. $4 \times 5 = 20, 5 \times 4 = 20$
5. $3 \times 6 = 18, 6 \times 3 = 18$
6. $4 \times 4 = 16, 2 \times 8 = 16$
7. $3 \times 5 = 15, 15 \div 3 = 5, 15 \div 5 = 3$
8. $4 \times 6 = 24, 24 \div 4 = 6, 24 \div 6 = 4$
9. $5 \times 7 = 35, 35 \div 5 = 7, 35 \div 7 = 5$
10. $2 \times 8 = 16, 16 \div 2 = 8, 16 \div 8 = 2$
11. $6 \times 3 = 18, 18 \div 6 = 3, 18 \div 3 = 6$
12. $10 \times 4 = 40, 40 \div 10 = 4, 40 \div 4 = 10$
13. $8 \times 5 = 40, 40 \div 8 = 5, 40 \div 5 = 8$
14. $5 \times 4 = 20, 20 \div 5 = 4, 20 \div 4 = 5$
15. $9 \times 10 = 90, 90 \div 9 = 10, 90 \div 10 = 9$

Rocket $2 \times 3 = 6, 3 \times 2 = 6$
$6 \div 2 = 3, 6 \div 3 = 2$
$3 \times 4 = 12, 4 \times 3 = 12$
$12 \div 3 = 4, 12 \div 4 = 3$
$2 \times 6 = 12, 6 \times 2 = 12$
$12 \div 2 = 6, 12 \div 6 = 2$

Page 22
Multiplying and dividing

1.

×	1	2	3	4	5	6
1	1	2	3	4	5	6
2	2	4	6	8	10	12
3	3	6	9	12	15	18
4	4	8	12	16	20	24
5	5	10	15	20	25	30
6	6	12	18	24	30	36

2. $8 \div 4 = 2$
   $8 \div 2 = 4$
2. Answers will vary
3. True
4. True
5. $48 \div 4 = 12$
6. $24 \times 6 = 144$
7. $120 \div 8 = 15$
8. $96 \div 4 = 24$
9. $96 \div 12 = 8$
10. $120 \div 15 = 8$
11. $15 \times 8 = 120$
12. $24 \times 4 = 96$
13. $144 \div 6 = 24$

Page 23
Multiplying and dividing by 2

1. $3 \times 2p = 6p$
2. $8 \times 2p = 16p$
3. $4 \times 2p = 8p$
4. $9 \times 2p = 18p$
5. $6 \times 2p = 12p$
6. $2 \times 2p = 4p$
7. $10 \times 2p = 20p$
8. $7 \times 2p = 14p$
9. $17 \times 2p = 34p$
10. True
11. True
12. False
13. 8p change
14. 9 pairs
15. 9
16. 24

Page 24
Multiplying and dividing by 4

1. 16
2. 24
3. $12 \div 4 = 3$
4. $20 \div 4 = 5$
5. $40 \div 4 = 10$
6. $4 \div 4 = 1$
7. $36 \div 4 = 9$
8. $28 \div 4 = 7$
9. $32 \div 4 = 8$
10. $24 \div 4 = 6$
11. $16 \div 4 = 4$
Rocket 44, 48, 52, 56, 60, 64, 68, 72, 76
12. 24 season changes
13. 56 slices
14. 9 tables with 2 chairs left over

Page 25
Multiplying and dividing by 8

1. 40
2. 24
3. 32
4. 48
5. 72
6. 56
7. $4 \times 8 = 32$
8. $6 \times 8 = 48$
9. $9 \times 8 = 72$
10. $3 \times 8 = 24$
11. $8 \times 8 = 64$
12. $2 \times 8 = 16$
Rocket 5 and 10 fingers show multiples of 10. 48 hands are needed to show 240.
13. $2 \times 8 = 16$
14. $7 \times 8 = 56$
15. $5 \times 8 = 40$
16. $32 \div 8 = 4$
17. $11 \times 8 = 88$
18. $64 \div 8 = 8$
19. $6 \times 8 = 48$
20. $9 \times 8 = 72$
21. $0 \times 8 = 0$
22. $24 \div 8 = 3$
23. $8 \div 8 = 1$
24. $20 \times 8 = 160$

Page 26
Twos, fours and eights

1. $2 \times 4 = 8 \times 1$
2. $4 \times 6 = 8 \times 3$
3. $10 \times 2 = 4 \times 5$
4. $32 \div 4 = 2 \times 4$
5. $16 \div 2 = 2 \times 4$
6. $36 \div 9 = 20 \div 5$
Rocket Answers will vary
7. 9p
8. 36p, 4p change
9. 10
10. 80
11. 16
12. Answers will vary but may include $8 \div 2 = 4$, $8 \div 4 = 2$, $4 \div 2 = 2$, $2 \times 2 = 4$,

Page 27
Fives and tens

1. $35 \div 5 = 7$
2. $80 \div 10 = 8$
3. $40 \div 5 = 8$
4. $60 \div 10 = 6$
5. $25 \div 5 = 5$
6. $70 \div 10 = 7$
7. $15 \div 5 = 3$
8. $40 \div 10 = 4$
9. $45 \div 5 = 9$
10. $5 \times 8 = 40$. Gill needs 8 coins to make £1.

11. $5 \times 9 = 45$. 9 teams can be made.
12. $9 \times 5 = 45$. 9 cars are needed.
13. $60 \div 5 = 12 \rightarrow 6 \div 10 = 6$
    double 6 = 12
14. $90 \div 5 = 18 \rightarrow 90 \div 10 = 9$
    double 9 = 18
15. $70 \div 5 = 14 \rightarrow 70 \div 10 = 7$
    double 7 = 14
16. $160 \div 5 = 32 \rightarrow 160 \div 10 = 16$
    double 16 = 32
17. $120 \div 5 = 24 \rightarrow 120 \div 10 = 12$
    double 12 = 24
18. $230 \div 5 = 46 \rightarrow 230 \div 10 = 23$
    double 23 = 46

Page 28
Fives and tens

1. $5 \times 8 = 4 \times 10$
2. $5 \times 4 = 2 \times 10$
3. $6 \times 3 = 3 \times 6$
4. $50 \div 10 = 5 \times 1$
5. $40 \div 5 = 80 \div 10$
6. $60 \div 5 = 120 \div 10$ (or any other multiple of 10)
Rocket Answers will vary.
7. $3 \times 10 = 30$
8. $5 \times 5 = 25$
9. $5 \times 8 = 40$
10. $10 \times 9 = 90$
11. $8 \times 10 = 80$
12. $7 \times 5 = 35$
13. $8 \times 10 = 80$
14. $5 \times 9 = 45$
15. $4 \times 10 = 40$
16. $4 \times 5 = 20$
17. Yes

Page 29
Threes and sixes

1. $4 \times 3 = 12$
   $4 \times 6 = 24$
2. $5 \times 3 = 15$
   $5 \times 6 = 30$
3. $8 \times 3 = 24$
   $8 \times 6 = 48$
4. $6 \times 3 = 18$
   $6 \times 6 = 36$
5. $9 \times 3 = 27$
   $9 \times 6 = 54$
6. 18
7. 24
8. 12
9. 36
10. $6 \times 3 = 18$
    $4 \times 1 = 4$
    18 + 4 = 22 points
11. $8 \times 3 = 24$
    $3 \times 1 = 3$
    24 + 3 = 27 points
12. $9 \times 3 = 27$
    $5 \times 1 = 5$
    27 + 5 = 32 points

13. 4 × 3 = 12
6 × 1 = 6
12 + 6 = 18 points
14. 7 × 3 = 21
4 × 1 = 4
21 + 4 = 25 points
15. 3 × 3 = 9
2 × 1 = 2
9 + 2 = 11 points
Rocket
10. 6 × 4 = 24
4 × 2 = 8
24 + 8 = 32 points
11. 8 × 4 = 32
3 × 2 = 6
32 + 6 = 38 points
12. 9 × 4 = 36
5 × 2 = 10
36 + 10 = 46 points
13. 4 × 4 = 16
6 × 2 = 12
16 + 12 = 28 points
14. 7 × 4 = 28
4 × 2 = 8
28 + 8 = 36 points
15. 3 × 4 = 12
2 × 2 = 4
12 + 4 = 16 points

Page 30
Nines

1. 2 × 9 = 18
2. 3 × 9 = 27
3. 72 ÷ 9 = 8
4. 90 ÷ 9 = 10
5. 4 × 9 = 36
6. 9 × 1 = 9
7. 81 ÷ 9 = 9
8. 45 ÷ 5 = 9
9. 8 × 9 = 72
10. 54 ÷ 6 = 9
11. 7 × 9 = 63
12. 36 ÷ 4 = 9
13. True
14. False
15. False
16. False
17. False
18. False
Rocket Answers will vary

Page 31
Nines

1. a: 18
b: 72
2. c: 36
d: 63
3. e: 90
f: 540
4. g: 270
h: 810
5. 81p
6. 8 dresses

7. 10 days with 4 cans left over
Rocket Answers will vary
Examples might include:
504, 513, 522, 531, 540;
603, 612, 621, 630;
702, 711, 720;
801, 810;
900, 1008, 2754, 8361…

Page 32
Threes, sixes and nines

1. 3 × 8 = 6 × 4
2. 9 × 2 = 6 × 3
3. 6 × 2 = 3 × 4
4. 36 ÷ 6 = 2 × 3
5. 18 ÷ 2 = 3 × 3
6. 36 ÷ 9 = 24 ÷ 6
Rocket Answers will vary.
7. 7 × 3 = 21p
8. 5 × 9 = 45p
9. 4 × 6 = 24p
10. 3 × 6 = 18p
11. 7 × 6 = 42p
12. 7 × 9 = 63p
13. 4 × 9 = 36p
14. 7 × 6 = 42p
15. 7 × 3 = 21p
16. 4 × 9 = 36p
17. 4 × 3 = 12p
18. 9 × 6 = 54p

Page 33
Multiplying by 10 and 100

1. 450 m
2. 380 m
3. 620 m
4. 740 m
5. 280 m
6. 1040 m
7. 960 m
8. 650 m
9. 190 m
After 100 laps:
1. 4500 m
2. 3800 m
3. 6200 m
4. 7400 m
5. 2800 m
6. 10 400 m
7. 9600 m
8. 6500 m
9. 1900 m
Rocket Hedgehogs 1, 3, 4, 6, 7 and 8
10. 35 × 10 = 350
11. 47 × 100 = 4700
12. 280 × 10 = 2800
13. 360 × 100 = 36 000
14. 7 × 100 = 700
15. 64 × 10 = 640
16. 68 × 100 = 6800
17. 11 × 100 = 1100
18. 49 × 10 = 490

Page 34
Multiplying by 10 and 100

1. 3 × 100 = 300
2. 5 × 100 = 500
3. 4 × 100 = 400
4. 8 × 100 = 800
5. 6 × 100 = 600
6. 7 × 100 = 700
7. 8 × 10 = 80 cm
8. 3 × 10 = 30 cm
9. 6 × 10 = 60 cm
10. 11 × 10 = 110 cm
11. 120 seconds is 2 minutes:
2 × 10 = 20 cm
12. 300 seconds is 5 minutes:
5 × 10 = 50 cm
Rocket 60 × 10 = 600 cm
13. 4 × 100 = 400
14. 7 × 10 = 70
15. 9 × 100 = 900
16. 6 × 10 = 60
17. 1 × 100 = 100
18. 2 × 10 = 20

Page 35
Multiplying

1. 3 × 100 = 300 cm
2. 7 × 100 = 700 cm
3. 5 × 100 = 500 cm
4. 9 × 100 = 900 cm
5. 6 × 100 = 600 cm
6. 2 × 100 = 200 cm
Rocket
1. 7 m
2. 3 m
3. 5 m
4. 1 m
5. 4 m
6. 8 m
7. 3 × 20 = 60
8. 4 × 40 = 160
9. 5 × 30 = 150
10. 4 × 50 = 200
11. 6 × 30 = 180
12. 3 × 40 = 120
Rocket 6 necklaces
Left over beads = 100 + 90 +
140 + 120 + 60 = 510

Page 36
Dividing by 10 and 100

1. 4800 ÷ 100 = £48
2. 750 ÷ 10 = £75
3. 6400 ÷ 100 = £64
4. 800 ÷ 100 = £8
5. 4600 ÷ 10 = £460
6. 770 ÷ 10 = £77
7. 690 ÷ 10 = £69
8. 9800 ÷ 10 = £980
9. 7400 ÷ 100 = £74

Rocket $10 \times 365 = 3650p = £36·50$
$3 \times 12 = £36$
10p a day for a year is better than £3 a month for a year.
10. $840 \div 10 = 84$
11. $5600 \div 100 = 56$
12. $7600 \div 10 = 760$
13. $7900 \div 100 = 79$
14. $4800 \div 100 = 48$
15. $8500 \div 100 = 85$
16. $950 \div 10 = 95$
17. $480 \div 10 = 48$
18. $8000 \div 10 = 800$

Page 37
Multiplying and dividing

1. $3 \times 20p = 60p$
2. $2 \times 30p = 60p$
3. $5 \times 40p = 200p = £2$
4. $7 \times 50p = 350p = £3·50$
5. $6 \times 20p = 120p = £1·20$
6. $4 \times 30p = 120p = £1·20$
7. $3 \times 40p = 120p = £1·20$
8. $9 \times 50p = 450p = £4·50$
9. $10 \times 40p + 4 \times 50p =$
$400p + 200p = 600p = £6·00$
10. $6 \times 30p + 5 \times 20p =$
$180p + 100p = 280p = £2·80$
Rocket 25 clown masks, 12 alien masks, 16 teddy bear masks, 10 monkey masks
11. $3 \times 30 = 90$
12. $5 \times 40 = 200$
13. $6 \times 30 = 180$
14. $8 \times 50 = 400$
15. $40 \times 3 = 120$
16. $4 \times 20 = 80$
17. $50 \times 5 = 250$
18. $9 \times 30 = 270$
19. $2 \times 60 = 120$

Page 38
Multiplication facts

1. $3 \times 6 = 18$ $30 \times 6 = 180$
$3 \times 60 = 180$ $30 \times 60 = 1800$
2. $4 \times 8 = 32$ $40 \times 8 = 320$
$4 \times 80 = 320$ $40 \times 80 = 3200$
3. $3 \times 7 = 21$ $30 \times 7 = 210$
$3 \times 70 = 210$ $30 \times 70 = 2100$
4. $4 \times 5 = 20$ $40 \times 5 = 200$
$4 \times 50 = 200$ $40 \times 50 = 2000$
5. $5 \times 7 = 35$ $50 \times 7 = 350$
$5 \times 70 = 350$ $50 \times 70 = 3500$
6. $4 \times 9 = 36$ $40 \times 9 = 360$
$4 \times 90 = 360$ $40 \times 90 = 3600$
7. $6 \times 5 = 30$ $60 \times 5 = 300$
$6 \times 50 = 300$ $60 \times 50 = 3000$
8. $6 \times 9 = 54$ $60 \times 9 = 540$
$6 \times 90 = 540$ $60 \times 90 = 5400$
9. $3 \times 4 = 12$ $30 \times 4 = 120$
$3 \times 40 = 120$ $30 \times 40 = 1200$
10. $8 \times 4 = 32$ $80 \times 4 = 32$
$8 \times 40 = 320$ $80 \times 40 = 3200$

11. $8 \times 7 = 56$ $80 \times 7 = 560$
$8 \times 70 = 560$ $80 \times 70 = 5600$
12. $8 \times 9 = 72$ $80 \times 9 = 720$
$8 \times 90 = 720$ $80 \times 90 = 7200$
Rocket Answers will vary but may include:
$24 \times 10 = 240$ $15 \times 16 = 240$
$48 \times 5 = 240$ $80 \times 3 = 240$
$120 \times 2 = 240$ $60 \times 4 = 240$
$30 \times 8 = 240$
13. $40 \times 50 = 2000$ pencils
14. $20 \times 70 = 1400$ rubbers
15. $80 \times 30 = 2400$ rulers
16. $90 \times 40 = 3600$ pencil sharpeners

Page 39
Multiplication facts

1. $5 \times 6 = 30$
$500 \times 6 = 3000$ $5 \times 600 = 3000$
2. $5 \times 8 = 40$
$500 \times 8 = 4000$ $5 \times 800 = 4000$
3. $5 \times 7 = 35$
$500 \times 7 = 3500$ $5 \times 700 = 3500$
4. $4 \times 4 = 16$
$400 \times 4 = 1600$ $4 \times 400 = 1600$
5. $4 \times 7 = 28$
$400 \times 7 = 2800$ $4 \times 700 = 2800$
6. $4 \times 9 = 36$
$400 \times 9 = 3600$ $4 \times 900 = 3600$
7. $6 \times 5 = 30$
$600 \times 5 = 3000$ $6 \times 500 = 3000$
8. $6 \times 7 = 42$
$600 \times 7 = 4200$ $6 \times 700 = 4200$
9. $6 \times 9 = 54$
$600 \times 9 = 5400$ $6 \times 900 = 5400$
10. $9 \times 5 = 45$
$900 \times 5 = 4500$ $9 \times 500 = 4500$
11. $9 \times 4 = 36$
$900 \times 4 = 3600$ $9 \times 400 = 3600$
12. $9 \times 9 = 81$
$900 \times 9 = 8100$ $9 \times 900 = 8100$
Rocket Answers will vary
13. £2000
14. $2100\,kg$
15. £80 a week for 60 weeks

Page 40
Multiplying

1. $33 \times 4 = 132$
2. $32 \times 5 = 160$
3. $21 \times 9 = 189$
4. $64 \times 5 = 320$
5. $47 \times 2 = 94$
6. $13 \times 5 = 65$
7. $48 \times 9 = 432$
8. $21 \times 4 = 84$
9. $15 \times 4 = 60$
10. $73 \times 4 = 292$
11. $40 \times 9 = 360$
12. $27 \times 9 = 243$
13. 15p
14. $150\,m$
15. 15p

16. 5
Rocket Answers will vary but might include:
$6 \times 2 \times 2 = 24$
$4 \times 2 \times 3 = 24$

Page 41
Doubling

1. $2 \times 10 = 20$ $2 \times 3 = 6$
$20 + 6 = 26$ $2 \times 13 = 26$
2. $2 \times 20 = 40$ $2 \times 4 = 8$
$40 + 8 = 48$ $2 \times 24 = 48$
3. $2 \times 30 = 60$ $2 \times 1 = 2$
$60 + 2 = 62$ $2 \times 31 = 62$
4. $2 \times 40 = 80$ $2 \times 2 = 4$
$80 + 4 = 84$ $2 \times 42 = 84$
5. $2 \times 50 = 100$ $2 \times 1 = 2$
$100 + 2 = 102$ $2 \times 51 = 102$
6. $2 \times 40 = 80$ $2 \times 7 = 14$
$80 + 14 = 94$ $2 \times 47 = 94$
7. $2 \times 10 = 20$ $2 \times 8 = 16$
$20 + 16 = 36$ $2 \times 18 = 36$
8. $2 \times 30 = 60$ $2 \times 6 = 12$
$60 + 12 = 72$ $2 \times 36 = 72$
9. $2 \times 50 = 100$ $2 \times 4 = 8$
$100 + 8 = 108$ $2 \times 54 = 108$
10. $2 \times 50 = 100$ $2 \times 8 = 16$
$100 + 16 = 116$ $2 \times 58 = 116$
11. $2 \times 60 = 120$ $2 \times 8 = 16$
$120 + 16 = 136$ $2 \times 68 = 136$
12. $2 \times 70 = 140$ $2 \times 8 = 16$
$140 + 16 = 156$ $2 \times 78 = 156$
Rocket $2 \times 75 = 150$,
so $4 \times 75 = 2 \times 150 = 300$,
and $8 \times 75 = 4 \times 150 = 600$
$16 \times 75 = 2 \times 600 = 1200$
13. 44p
14. 108 g of butter and 134 g of cheese.
15. £114

Page 42
Multiplying

1. 78 coins
2. £260
Rocket
3. Multiply by 10 and then by 5
4. Multiply by 100 and then double it
5. Multiply by 10 and then add the number
6. Multiply by 10 and then subtract the number
$43 \times 50 = 2150$
$43 \times 200 = 8600$
$43 \times 11 = 473$
$43 \times 9 = 387$
$18 \times 50 = 900$
$18 \times 200 = 3600$
$18 \times 11 = 198$
$18 \times 9 = 162$
$160 \times 50 = 8000$
$160 \times 200 = 32\,000$
$160 \times 11 = 1760$
$160 \times 9 = 1440$

7. $73 \times 10 = 730$
8. $42 \times 100 = 4200$
9. $39 \times 100 = 3900$
10. $860 \times 10 = 8600$
11. $47 \times 5 = 235$
12. $96 \times 20 = 1920$
13. $28 \times 50 = 1400$
14. $86 \times 20 = 1720$
15. $32 \times 200 = 6400$
16. $38 \times 50 = 1900$

Page 43
Dividing

1. $20 \div 5 = 4$
   $20 \div 4 = 5$
2. $24 \div 6 = 4$
   $24 \div 4 = 6$
3. $30 \div 6 = 5$
   $30 \div 5 = 6$
4. $21 \div 3 = 7$
   $21 \div 7 = 3$
5. $12 \div 6 = 2$
   $12 \div 2 = 6$
6. $28 \div 7 = 4$
   $28 \div 4 = 7$

Rocket Answers will vary. Possible answers include:

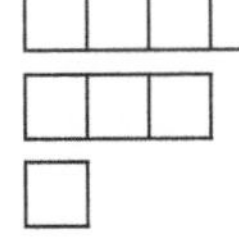

7. $27 \div 3 = 9$ teams exactly
8. $27 \div 4 = 6$ teams, 3 children left over
9. $27 \div 2 = 13$ teams, 1 child left over
10. $27 \div 10 = 2$ teams, 7 children left over
11. $27 \div 5 = 5$ teams, 2 children left over
12. $27 \div 6 = 4$ teams, 3 children left over

Rocket $24 \div 2 = 12$ teams, $24 \div 3 = 8$ teams, $24 \div 4 = 6$ teams, $24 \div 6 = 4$ teams, $24 \div 8 = 3$ teams, $24 \div 12 = 2$ teams

Page 44
Dividing

1. $15 \div 4 = 3 \text{ r } 3$
2. $29 \div 4 = 7 \text{ r } 1$
3. $38 \div 4 = 9 \text{ r } 2$
4. $22 \div 4 = 5 \text{ r } 2$
5. $23 \div 5 = 4 \text{ r } 3$
6. $18 \div 5 = 3 \text{ r } 3$
7. $44 \div 5 = 8 \text{ r } 4$
8. $36 \div 5 = 7 \text{ r } 1$

Rocket Answers will vary. Possible answers include 6 and 18 in the 2× and 3× tables, or 42 and 84 in the 6× and 7× tables.

9. $13 \div 3 = 4 \text{ r } 1$
   5 chair lifts
10. $22 \div 3 = 7 \text{ r } 1$
    8 chair lifts
11. $28 \div 3 = 9 \text{ r } 1$
    10 chair lifts
12. $21 \div 4 = 5 \text{ r } 1$
    6 cable cars
13. $43 \div 4 = 10 \text{ r } 3$
    11 cable cars
14. $82 \div 4 = 20 \text{ r } 2$
    21 cable cars
15. $17 \div 5 = 3 \text{ r } 2$
    4 toboggans
16. $28 \div 5 = 5 \text{ r } 3$
    6 toboggans
17. $61 \div 5 = 12 \text{ r } 1$
    13 toboggans

Rocket 60 people or any multiple of 60.

Page 45
Dividing

1. true
2. true
3. false $36 \div 10 = 3 \text{ r } 6$
4. false $25 \div 3 = 8 \text{ r } 1$
5. true
6. false $38 \div 4 = 9 \text{ r } 2$
7. true
8. false $43 \div 5 = 8 \text{ r } 3$
9. true
10. £1·80
11. 9 piles of 4 bricks, 1 pile of 2 bricks
12. 4 days with 1 slice left over
13. 38p

Rocket Answers will vary.
Numbers that give a remainder of 1 when divided by 2 and 3 include: 13, 19, 25, 31, 37, 43, etc.
Numbers that give a remainder of 1 when divided by 2 and 5 include: 11, 21, 31, 41, etc.

Page 46
Dividing 2-digit numbers by 5

1. $89 \div 5 = 17 \text{ r } 4$
2. $76 \div 5 = 15 \text{ r } 1$
3. $54 \div 5 = 10 \text{ r } 4$
4. $66 \div 5 = 13 \text{ r } 1$
5. $94 \div 5 = 18 \text{ r } 4$
6. $99 \div 5 = 19 \text{ r } 4$
7. $102 \div 5 = 20 \text{ r } 2$
8. $127 \div 5 = 25 \text{ r } 2$

Rocket Answers will vary.
9. Children's answers may vary, but the first possible answer above 50 is 63.

Page 47
Multiplying

1. £3.12
2. £2.40. 55p cheaper
3. £190

Rocket Answers will vary
4. $5 \times 43 = 215$
5. $7 \times 26 = 182$
6. $6 \times 48 = 288$
7. $5 \times 28 = 140$

Rocket Answers will vary

Page 48
Multiplying and dividing

1. $6 \times 37 = £222$
2. $8 \times 64 = £512$
3. $2 \times 42 = £84$
4. $3 \times 36 = £108$
5. £4
6. £42
7. 14 ping pong balls with 2 left over.

Rocket Answers will vary.

Multiplication and Division PPMs

PPM 102
2s, 3s, 4s ...
1. 12
2. 10
3. 15
4. 24
5. 20
6. 18

PPM 103
Making towers of 4

1. 4
2. 6
3. 3
4. 7
5. 2
6. 5

PPM 104

Arrays and number lines

1. 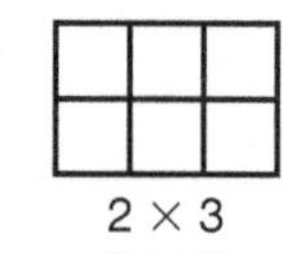 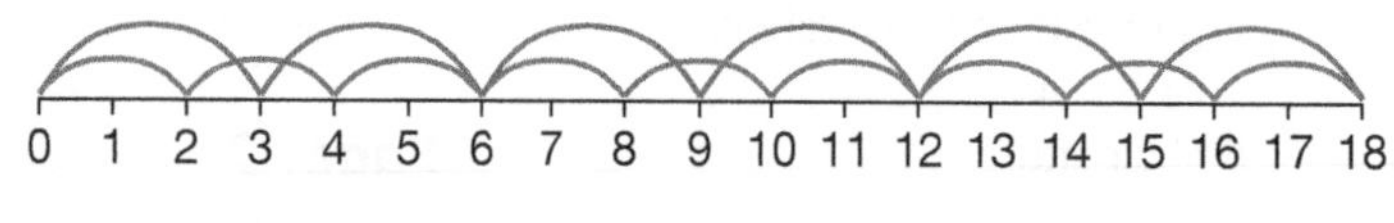

2×3
3×2

2. 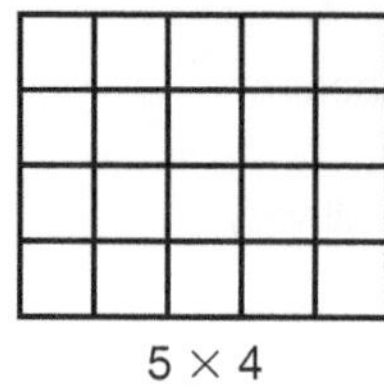 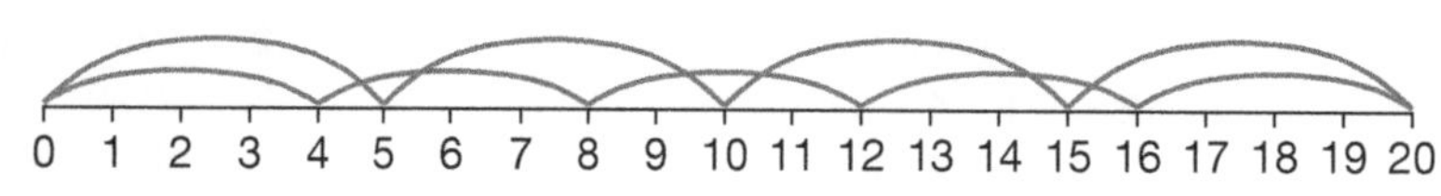

5×4
4×5

3. 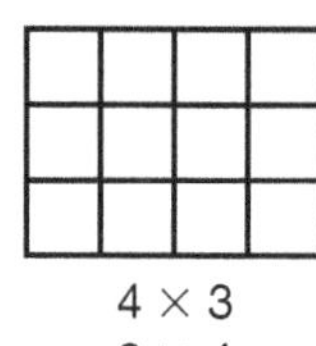 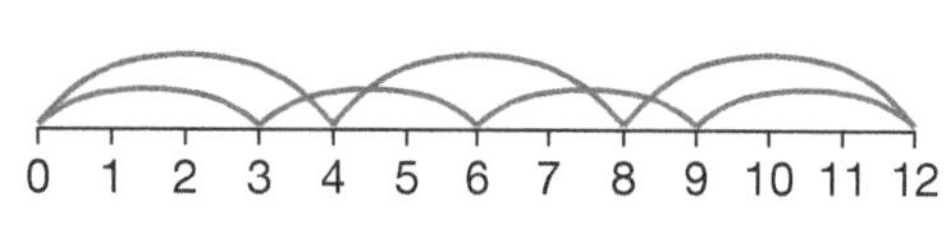

4×3
3×4

4. 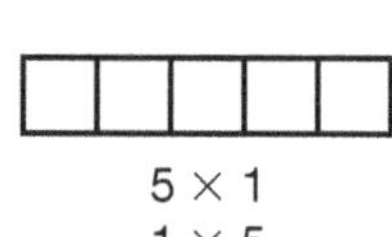 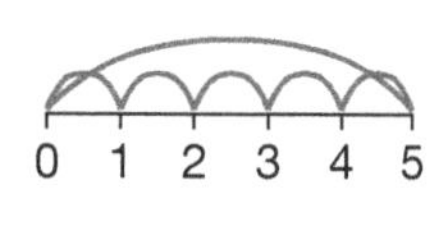

5×1
1×5

PPM 105

Arrays and number lines

1. $2 \times 4 = 8$ $\quad$ $4 \times 2 = 8$

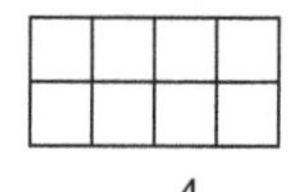
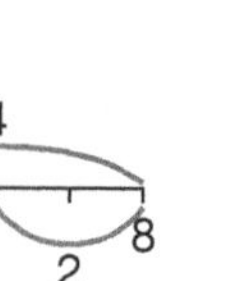

2. $4 \times 6 = 24$ $\quad$ $6 \times 4 = 24$

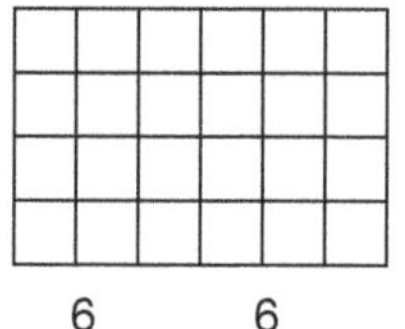
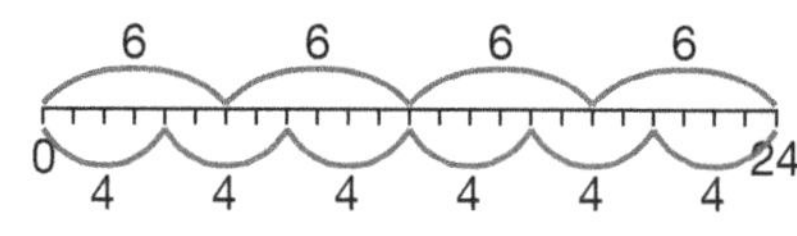

3. $4 \times 5 = 20$ $\quad$ $5 \times 4 = 20$

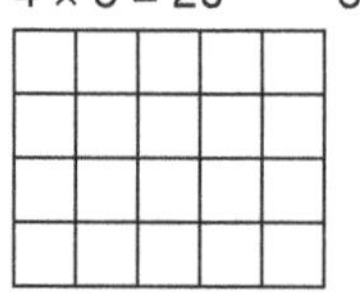
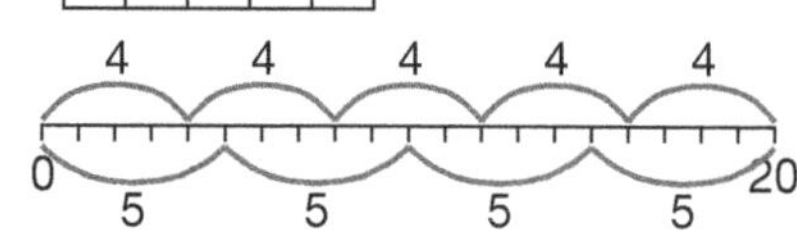

4. $3 \times 2 = 6$ $\quad$ $2 \times 3 = 6$

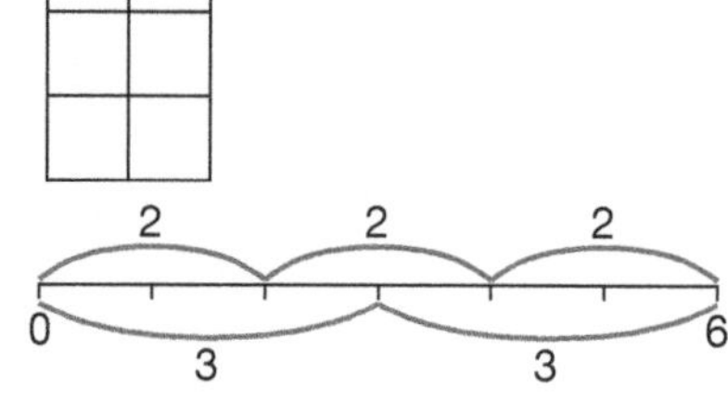

5. $4 \times 7 = 28$ $\quad$ $7 \times 4 = 28$

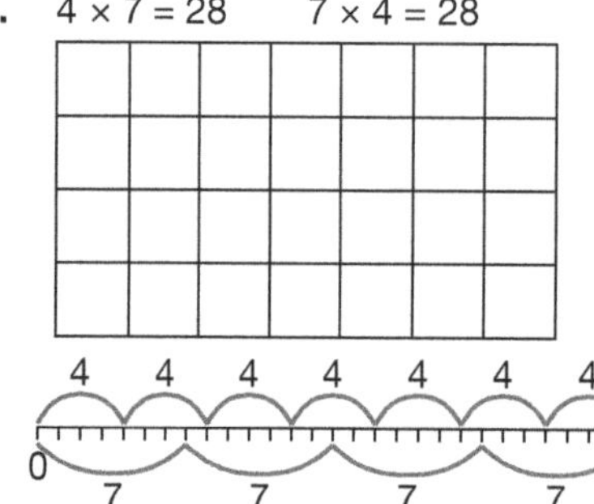

PPM 106

Arrays and number lines

1. 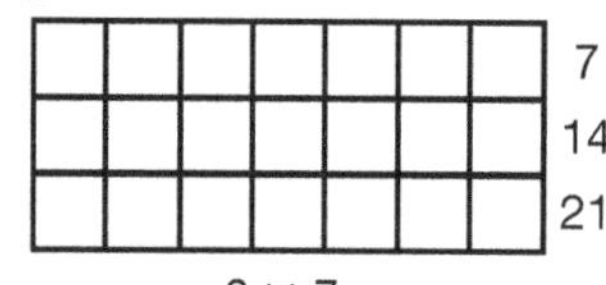

7
14
21

3×7

2. 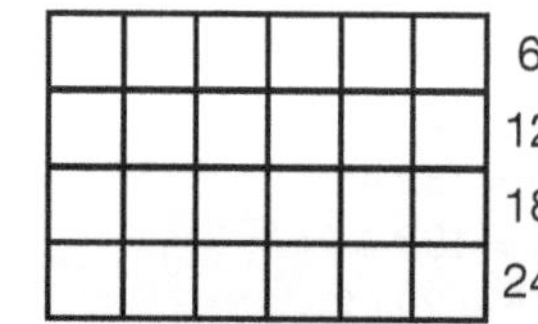

2
4
6
8

4×2

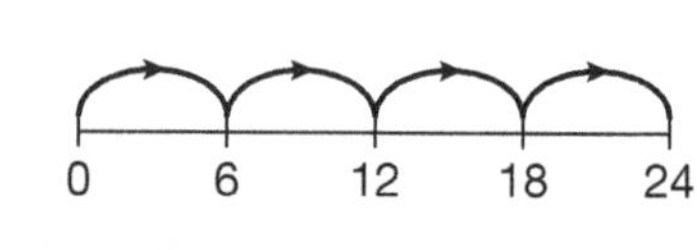

3. 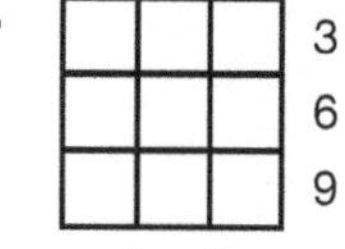

6
12
18
24

4×6

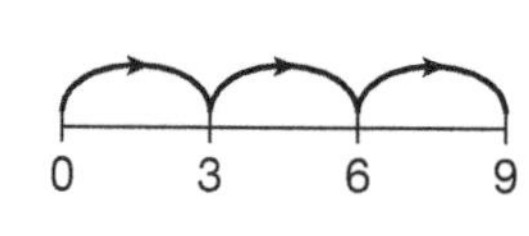

4.

3
6
9

3×3

PPM 107

Fours

1. 8
2. 16
3. 20
4. 40
5. 6
6. 12
7. 10
8. 20
9. 14
10. 28
11. 18
12. 36
13. 22
14. 44

PPM 108

Eights

1. 8
2. 16
3. 24
4. 32
5. 40
6. 48
7. 56
8. 64
9. 72
10. 80

PPM 109

Eights

1.

4s	4	8	12	16	20	24	28	32	36	40
8s	8	16	24	32	40	48	56	64	72	80

2. $5 \times 8 = 40$
3. $4 \times 8 = 32$
4. $9 \times 8 = 72$
5. $7 \times 8 = 56$
6. $2 \times 8 = 16$
7. $12 \times 8 = 96$
8.

2s	2	4	6	8	10	12	14	16	18	20
4s	4	8	12	16	20	24	28	32	36	40
8s	8	16	24	32	40	48	56	64	72	80

9.

4s	4	(8)	12	(16)	20	(24)	28	(32)	36	(40)
8s	(8)	(16)	(24)	(32)	(40)	(48)	56	(64)	72	(80)
16s	(16)	(32)	(48)	(64)	(80)	96	112	128	144	160

32 is in the 4s, the 8s and the 16s.

PPM 110

Fives

1. $6 \times 5 = 30$
2. $4 \times 5 = 20$
3. $3 \times 5 = 15$
4. $7 \times 5 = 35$
5. $8 \times 5 = 40$
6. $5 \times 5 = 25$

PPM 111

Fives and tens

1.

5s	5	10	15	20	25	30	35	40	45	50

2. $5 \times 5 = 25$
3. $4 \times 5 = 20$
4. $9 \times 5 = 45$
5. $7 \times 5 = 35$
6. $2 \times 5 = 10$
7. $12 \times 5 = 60$
8.

10s	10	20	30	40	50	60	70	80	90	100

9. $4 \times 10 = 40$
10. $9 \times 10 = 90$
11. $3 \times 10 = 30$
12. $7 \times 10 = 70$
13. $5 \times 10 = 50$
14. $2 \times 10 = 20$
15.

5s	5	10	15	20	25	30	35	40	45	50
10s	10	20	30	40	50	60	70	80	90	100

Answers will vary, but children should notice that every second number in the 5s row appears in the tens row.

PPM 112

Fives

1. 40
2. 20
3. 60
4. 30
5. 30
6. 15
7. 50
8. 25
9. 70
10. 35
11. 90
12. 45
13. 80
14. 40
15. 0
16. 0

PPM 113

Counting in sixes

1. Missing numbers: 24, 36, 42, 54, 60, 72

2.

$1 \times 3 = 3$	$1 \times 6 = 6$
$2 \times 3 = 6$	$2 \times 6 = 12$
$3 \times 3 = 9$	$3 \times 6 = 18$
$4 \times 3 = 12$	$4 \times 6 = 24$
$5 \times 3 = 15$	$5 \times 6 = 30$
$6 \times 3 = 18$	$6 \times 6 = 36$
$7 \times 3 = 21$	$7 \times 6 = 42$
$8 \times 3 = 24$	$8 \times 6 = 48$
$9 \times 3 = 27$	$9 \times 6 = 54$
$10 \times 3 = 30$	$10 \times 6 = 60$

PPM 114

Nines

1. $4 \times 9 = 36$
2. $6 \times 9 = 54$
3. $3 \times 9 = 27$
4. $1 \times 9 = 9$
5. $5 \times 9 = 45$
6. $2 \times 9 = 18$
7. $9 \times 9 = 81$
8. $7 \times 9 = 63$
9. $8 \times 9 = 72$
10. $1 \times 9 = 9$
11. $6 \times 9 = 54$
12. $45 \div 9 = 5$
13. $3 \times 9 = 27$
14. $5 \times 9 = 45$

15. $54 \div 9 = 6$
16. $4 \times 9 = 36$
17. $63 \div 9 = 7$
18. $9 \times 9 = 81$

PPM 115

Nines digits

	×9 table	Digit total	Digit difference
1.	18	9	7
2.	27	9	5
3.	36	9	3
4.	45	9	1
5.	54	9	1
6.	63	9	3
7.	72	9	5
8.	81	9	7
9.	90	9	9

PPM 116

Nines digits

1. 36
2. 72
3. 81
4. 45
5. 90
6. 54
7. 63
8. 27
9. 18

PPM 117

Multiplying and dividing

1. $3 \times 6 = 18, 18 \div 6 = 3$
2. $4 \times 4 = 16, 16 \div 4 = 4$
3. $2 \times 8 = 16, 16 \div 8 = 2$
4. $4 \times 7 = 28, 28 \div 7 = 4$
5. $5 \times 3 = 15, 15 \div 3 = 5$
6. $3 \times 7 = 21, 21 \div 7 = 3$
7. $3 \times 5 = 15, 15 \div 5 = 3$

PPM 118

Multiplying and dividing

1. 3
2. 4
3. 2
4. 4
5. 4
6. 7
7. 16
8. 3
9. 10
10. 2

PPM 119

Twos

1. 4
2. 8
3. 9
4. 10
5. 2
6. 5
7. 3
8. 1
9. 6
10. 7
11. 22
12. 24
13. 50

PPM 120

Fours

1. 16, 8, 24
    12, 32, 4
    28, 20, 36
2. 8, 36, 16
    32, 20, 40
    24, 12, 28
3. 24, 40, 12
    16, 32, 28
    36, 44, 20

PPM 121

Eights

1. 16
2. 56
3. 10
4. 5
5. 72
6. 64
7. 3
8. 2
9. 7
10. 24
11. 4
12. 80
13. 6
14. 40
15. 32
16. 9
17. 48
18. 8

PPM 122

Tens

Children should colour the pairs:
1, 10
2, 20
3, 30
4, 40
5, 50
6, 60
7, 70
8, 80
9, 90
10, 100

PPM 123

5s and 10s

Left-hand column answers
1. 3
2. 6
3. 5
4. 1
5. 9
6. 0
7. 7
8. 10
9. 2
10. 4
11. 8

Right-hand column answers
12. 2
13. 5
14. 9
15. 1
16. 7
17. 10
18. 4
19. 8
20. 3
21. 0
22. 6

PPM 124

Dividing by 5 and 10

Game determined by dice.

PPM 125

Threes

1. 12, 24, 6
    3, 15, 21
    18, 27, 9
2. 9, 21, 15
    18, 24, 6
    30, 12, 27
3. 36, 24, 12
    27, 9, 33
    15, 30, 21

PPM 126

Multiplying and dividing by 6

1. 30
2. 18
3. 10
4. 5
5. 42
6. 4
7. 36
8. 3
9. 24
10. 54
11. 8
12. 6
13. 12
14. 9
15. 48
16. 60
17. 7
18. 2

PPM 127

Multiplying and dividing by 9

1. 45
2. 27
3. 10
4. 2
5. 63
6. 5
7. 72
8. 6
9. 9
10. 90
11. 3
12. 18
13. 54
14. 4
15. 81
16. 8
17. 7
18. 36

PPM 128

Multiplying by 10, 100

1. 90
2. 70
3. 30
4. 180
5. 160
6. 120
7. 560
8. 800
9. 260
10. 1600
11. 2400
12. 3600
13. 9100
14. 800

PPM 129

Multiplying and dividing by 10 and 100

1. 70
2. 310
3. 500
4. 1700
5. 420
6. 3600
7. 1700
8. 6000
9. 110
10. 1200
11. 40
12. 27
13. 30
14. 18
15. 68
16. 9
17. 43
18. 270

PPM 130

Multiplying

1. 120
2. 120
3. 60
4. 100
5. 80
6. 160
7. 150
8. 180
9. 210
10. 240
11. 150
12. 240
13. 200
14. 140
15. 270
16. 280
17. 320
18. 280
19. 180
20. 360

PPM 131

Multiplying

Answers will vary

PPM 132

Multiplying

1. 56 – 560
2. 40 – 400
3. 72 – 720
4. 320
5. 640
6. 2400
7. 4800
8. 1600
9. 5600
10. 4000
11. 7200
12. 320
13. 560

PPM 133

Multiplying by 4 and 5

	×2	×4	×10	×5
16	32	64	160	80
22	44	88	220	110
17	34	68	170	85
24	48	96	240	120
26	52	104	260	130
19	38	76	190	95
25	50	100	250	125
36	72	144	360	180
41	82	164	410	205
54	108	216	540	270
57	114	228	570	285
69	138	276	690	345
97	194	388	970	485

PPM 134

Multiplying by 9

	×10	×9	
16	160	160 – 16 or 90 + 54	154
28	280	280 – 28 or 90 + 90 + 72	252
19	190	190 – 19 or 90 + 81	171
25	250	250 – 25 or 90 + 90 + 45	225
23	230	230 – 23 or 90 + 90 + 27	207
18	180	180 – 18 or 9072	162
29	290	290 – 29 or 90 + 90 + 81	261
36	360	360 – 36 or 90 + 90 + 90 + 54	324
41	410	410 – 41 or 90 + 90 + 90 + 90 + 9	369
54	540	540 – 54 or 90 + 90 + 90 + 90 + 90 + 36	486
57	570	570 – 57 or 90 + 90 + 90 + 90 + 90 + 63	513
69	690	690 – 69 or 90 + 90 + 90 + 90 + 90 + 90 + 81	621
97	970	970 – 97 or 90 + 90 + 90 + 90 + 90 + 90 + 90 + 90 + 63	873
99	990	990 – 99 or 90 + 90 + 90 + 90 + 90 + 90 + 90 + 90 + 81	891

PPM 135

Dividing by splitting

Find half of	Split and divide by 2	Answer
46	40 ÷ 2 = 20 6 ÷ 2 = 3	20 + 3 = 23
86	80 ÷ 2 = 40 6 ÷ 2 = 3	40 + 3 = 43
38	30 ÷ 2 = 15 8 ÷ 2 = 4	15 + 4 = 19
52	50 ÷ 2 = 25 2 ÷ 2 = 1	25 + 1 = 26
76	70 ÷ 2 = 35 6 ÷ 2 = 3	35 + 3 = 38
89	80 ÷ 2 = 40 9 ÷ 2 = 4 r1	44 r1
45	40 ÷ 2 = 20 5 ÷ 2 = 2 r1	22 r1
79	70 ÷ 2 = 35 9 ÷ 2 = 4 r1	39 r1
63	60 ÷ 2 = 30 3 ÷ 2 = 1 r1	31 r1
59	50 ÷ 2 = 25 9 ÷ 2 = 4 r1	29 r1

PPM 136
Dividing by 4

48	24	12
64	32	16
84	42	21
28	14	7
52	26	13
76	38	19
32	16	8
60	30	15
72	36	18
92	46	23

Multiplication and Division APMs

APM 312
Strides

1. 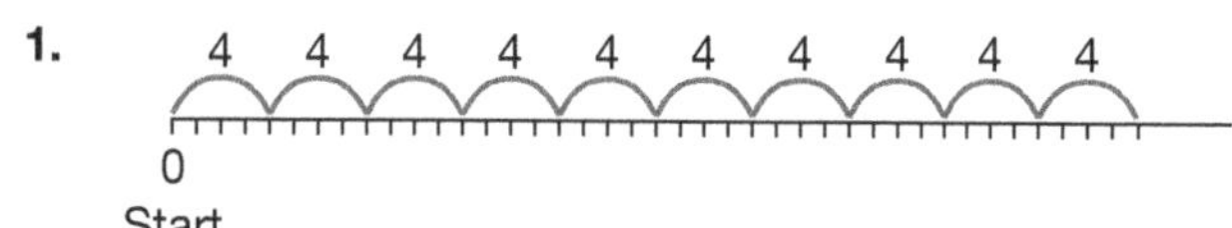
2. 40 units
3. 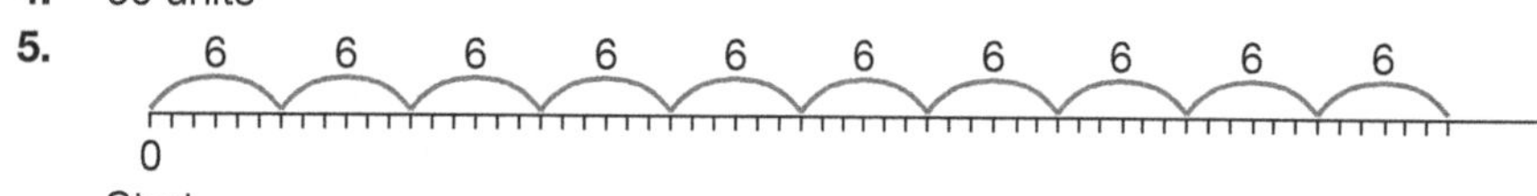
4. 50 units
5.
6. 60 units
7. Snowman One: 400 units,
   Snowman Two: 500 units,
   Snowman Three: 600 units
8. The snowmen's steps will line up
   after 120 units (4 × 5 × 6)

APM 315
What's in the square?

There are several correct answers, for
example:

4× table	20	36	40	24
2× table	10	18	20	12
3× table	15	27	30	18
8× table	40	72	80	48
	5× table	9× table	10× table	6× table

Fractions Pupil Book 4

Page 3
Tenths

1. Shapes e, h, j and l show tenths.
   Rocket Answers will vary.
2. Divide the shape into 10 equal
   parts instead of 5 equal parts. For
   example

3. Make each tenth the same size.
   For example

Page 4
Tenths

1. $\frac{3}{10}$ shaded. $\frac{7}{10}$ to be coloured in.
2. $\frac{7}{10}$ shaded. $\frac{3}{10}$ to be coloured in.
3. $\frac{4}{10}$ shaded. $\frac{6}{10}$ to be coloured in.
4. $\frac{2}{10}$ shaded. $\frac{8}{10}$ to be coloured in.
5. $\frac{9}{10}$ shaded. $\frac{1}{10}$ to be coloured in.
6. $\frac{5}{10}$ shaded. $\frac{5}{10}$ to be coloured in.
7. $\frac{10}{10}$ shaded. None to be coloured in.
8. $\frac{6}{10}$ shaded. $\frac{4}{10}$ to be coloured in.

Rocket It is written $\frac{1}{10}$ because this shows
that it is 1 part out of 10 parts.

9. $\frac{1}{10}$.
10. $\frac{2}{10}$

Page 5
Tenths

1. $\frac{1}{10}$ of 20 = 2
2. $\frac{1}{10}$ of 30 = 3
3. $\frac{1}{10}$ of 40 = 4
4. $\frac{1}{10}$ of 30 = 3
5. $\frac{1}{10}$ of 20 = 2
6. $\frac{1}{10}$ of 50 = 5

Rocket Children will find that it is not
possible to find a tenth of the
class if the amount of children in
the class is not in the 10-times
table.

7. $\frac{1}{10}$ of 60 = 6
8. $\frac{1}{10}$ of 80 = 8
9. $\frac{1}{10}$ of 100 = 10
10. $\frac{1}{10}$ of 130 = 13
11. $\frac{1}{10}$ of 200 = 20
12. $\frac{1}{10}$ of 170 = 17
13. $\frac{1}{10}$ of 440 = 44
14. $\frac{1}{10}$ of 320 = 32
15. $\frac{1}{10}$ of 510 = 51

Page 6
Tenths

1. 40 cubes
2. 70 cubes
3. 80 cubes
4. 120 cubes
5. 90 cubes
6. 130 cubes
7. 210 cubes
8. 320 cubes

Rocket Answers will vary and may
include words such as decagon
(a 10-sided shape) and
December (the 10th month in
the Roman calendar).

9. $\frac{1}{2}$ of 20 > $\frac{1}{10}$ of 50
10. $\frac{1}{2}$ of 50 > $\frac{1}{10}$ of 40
11. $\frac{1}{2}$ of 20 = $\frac{1}{10}$ of 100
12. $\frac{1}{2}$ of 30 < $\frac{1}{10}$ of 400
13. $\frac{1}{10}$ of £240 = £24

   $\frac{1}{2}$ of £46 = £23

   Kaya should choose $\frac{1}{10}$ of £240 =
   £24

Page 7
Tenths on a number line

1. a $\frac{1}{10}$, b $\frac{6}{10}$, c $1\frac{4}{10}$, d $1\frac{9}{10}$
2. e $2\frac{3}{10}$, f $2\frac{7}{10}$, g $2\frac{9}{10}$, h $3\frac{3}{10}$, i $3\frac{7}{10}$
3. j $4\frac{2}{10}$, k $4\frac{7}{10}$, l $4\frac{9}{10}$, m $5\frac{5}{10}$, n $5\frac{7}{10}$,
   o $5\frac{9}{10}$
4. p $9\frac{3}{10}$, q $9\frac{5}{10}$, r $9\frac{7}{10}$, s 10, t $10\frac{2}{10}$,
   u $10\frac{5}{10}$, v $10\frac{8}{10}$

5.–12.

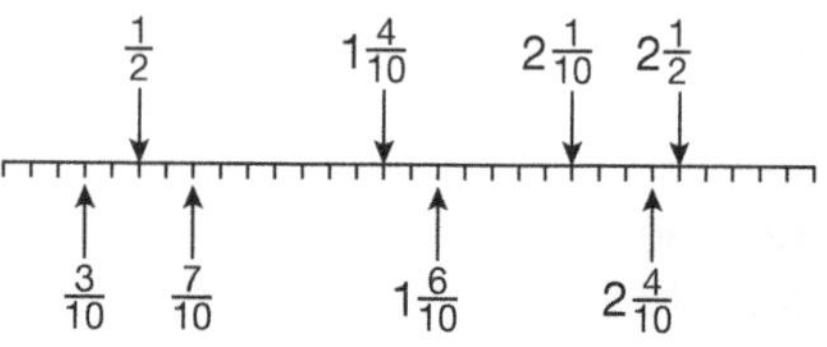

Page 8

Writing tenths

1. Answers will vary but answers may include
   Seven tenths are 7 parts of a whole object that is divided into 10 equal parts.

2. 40 km

Rocket Answers will vary but should show a representation of $\frac{2}{10}$, for example

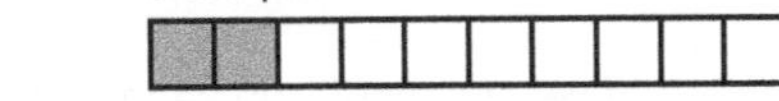

3. $\frac{8}{10}$
4. 160 km
5. $\frac{7}{10}$

Page 9

Fifths

1. Shapes b, e, h and k show fifths.

Rocket Answers will vary.

2. Make 5 equal parts instead of 4 parts, for example

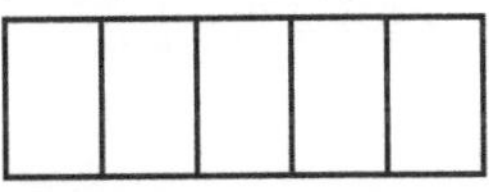

3. Make all 5 parts equal in size, for example

Page 10

Fifths

1. $\frac{1}{5}$
2. $\frac{2}{5}$
3. $\frac{0}{5}$
4. $\frac{3}{5}$
5. $\frac{1}{5}$
6. $\frac{4}{5}$
7. $\frac{5}{5}$
8. $\frac{2}{5}$
9. $\frac{3}{5}$
10. $\frac{3}{5}$
11. $\frac{1}{5}$
12. $\frac{4}{5}$
13. $\frac{2}{5}$
14. $\frac{1}{5}$
15. $\frac{2}{5}$
16. $\frac{2}{5}$

Page 11

Fifths and tenths

1. 25 cubes
2. 15 cubes
3. 40 cubes
4. 30 cubes
5. 45 cubes
6. 85 cubes
7. 50 cubes
8. 35 cubes

Rocket Answers will vary.

9. $\frac{1}{10}$ of 100 > $\frac{1}{5}$ of 40
10. $\frac{1}{10}$ of 40 < $\frac{1}{5}$ of 50
11. $\frac{1}{5}$ of 30 < $\frac{1}{10}$ of 100
12. $\frac{1}{10}$ of 20 = $\frac{1}{5}$ of 10
13. $\frac{1}{10}$ of £80 = £8
    $\frac{1}{5}$ of £40 = £8
    Laila should say that they are both the same.

Page 12

Fifths

1. 6 children
2. $\frac{2}{5}$ 12 children
3. 60 stickers
4. 16 games

Rocket Answers will vary but may include

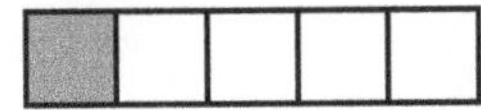

One fifth is a fraction written $\frac{1}{5}$. The numerator is 1 and the denominator is 5 which shows that it is 1 part out of 5.
One fifth is 1 part of a whole object that is divided into 5 equal parts.

Page 13

Fifths on a number line

1. a) $\frac{1}{5}$ b) $\frac{3}{5}$ c) $1\frac{2}{5}$
   d) 2   e) $2\frac{4}{5}$
2. f) $3\frac{3}{5}$ g) $4\frac{1}{5}$ h) 5
   i) $5\frac{4}{5}$ j) 6
3. k) $7\frac{2}{5}$ l) $8\frac{4}{5}$ m) $9\frac{1}{5}$
   n) $9\frac{3}{5}$ o) 10
4. p) 5   q) $5\frac{2}{5}$ r) 6
   s) $6\frac{3}{5}$ t) $9\frac{3}{5}$ u) 10
   v) $10\frac{2}{5}$

Rocket

0 1 2 3 4 5 6 7 8 9 10 11 12

There are 60 fifths between 0 and 12.
There are 500 fifths between 0 and 100 because there are 5 fifths in every 1.

Page 14

Hundredths

1. $\frac{35}{100}$
2. $\frac{15}{100}$
3. $\frac{85}{100}$
4. $\frac{21}{100}$
5. $\frac{70}{100}$

Page 15

Hundredths

1. 20 hundredths shaded
   80 hundredths not shaded
2. 40 hundredths shaded
   60 hundredths not shaded
3. 50 hundredths shaded
   50 hundredths not shaded
4. 34 hundredths shaded
   66 hundredths not shaded
5. 63 hundredths shaded
   37 hundredths not shaded
6. 75 hundredths shaded
   25 hundredths not shaded
7. 39 hundredths shaded
   61 hundredths not shaded
8. 92 hundredths shaded
   8 hundredths not shaded
9. 24 hundredths shaded
   76 hundredths not shaded
10. 100 hundredths
11. 50 hundredths

Rocket Answers will vary but may include weighing scales measured in kg and g, a measuring jug measured in m*l* and *l* and a ruler measured in mm and cm.

12. $\frac{20}{100} < \frac{40}{100}$
13. $\frac{63}{100} > \frac{34}{100}$
14. $\frac{24}{100} > \frac{20}{100}$
15. 1 whole = $\frac{100}{100}$
16. $\frac{39}{100} < \frac{75}{100}$
17. $\frac{1}{2} = \frac{50}{100}$

Page 16

Comparing fractions

1. $\frac{1}{2}$
2. $\frac{3}{4}$
3. 1 whole
4. They are the same.
5. a) $\frac{1}{4}$ b) $\frac{3}{4}$

Rocket Answers will vary, for example
$\frac{1}{2} = \frac{2}{4}$
$\frac{1}{4} < \frac{1}{2}$
$\frac{1}{2} > \frac{1}{4}$
$\frac{3}{4} > \frac{1}{2}$
etc.

Page 6

6. $\frac{49}{100}$
7. Shape c shows $\frac{75}{100}$

Rocket Answers will vary

Page 17

Comparing fractions

1. $\frac{1}{5}$
2. $\frac{1}{2}$
3. $\frac{6}{10}$
4. $\frac{7}{10}$
5. 1 whole

Example **6.** $\frac{3}{10} < \frac{3}{5}$

7. $\frac{1}{2} < \frac{3}{5}$

8. $\frac{4}{5} = \frac{8}{10}$

9. $\frac{7}{10} < \frac{4}{5}$

10. 1 whole $= \frac{2}{2}$

11. $\frac{1}{2}$ $\frac{1}{5}$ $\frac{1}{10}$

12. $\frac{3}{5}$ $\frac{1}{2}$ $\frac{2}{10}$

13. $\frac{4}{5}$ $\frac{1}{2}$ $\frac{2}{10}$

14. 1 whole $\frac{4}{5}$ $\frac{2}{10}$

Rocket $\frac{1}{5}$ $\frac{2}{10}$ $\frac{1}{2}$ $\frac{5}{10}$ $\frac{3}{5}$ $\frac{6}{10}$

Children will notice that $\frac{1}{5} = \frac{2}{10}$, $\frac{1}{2} = \frac{5}{10}$ and $\frac{3}{5} = \frac{6}{10}$.

Page 18

Comparing fractions

1. a) $\frac{2}{5}$ b) $\frac{3}{5}$ c) $\frac{4}{5}$
   d) $\frac{4}{10}$ e) $\frac{6}{10}$ f) $\frac{8}{10}$

2. $\frac{7}{10}$

3. $\frac{3}{10}$, $\frac{4}{10}$ or $\frac{2}{5}$

4. $\frac{1}{10}$ $\frac{1}{5}$ $\frac{1}{2}$

5. $\frac{1}{5}$ $\frac{4}{10}$ $\frac{3}{5}$ $\frac{7}{10}$

Rocket Answers will vary but may include

$\frac{1}{2} = \frac{5}{10}$

$\frac{2}{5} = \frac{4}{10}$

$\frac{8}{10} = \frac{4}{5}$

$\frac{5}{5} = \frac{2}{2}$

Page 19

Ordering fractions

1. $\frac{1}{2} > \frac{1}{5}$

2. $\frac{6}{10} > \frac{2}{4}$

3. $\frac{3}{5} < \frac{6}{8}$

4. $\frac{2}{5} < \frac{3}{4}$

5. $\frac{2}{4} = \frac{5}{10}$

6. $\frac{7}{10} < \frac{3}{4}$

Rocket

1. $\frac{1}{2}$ $\frac{4}{5}$

2. $\frac{4}{10}$ $\frac{2}{4}$

3. $\frac{2}{5}$ $\frac{2}{8}$

4. $\frac{3}{5}$ $\frac{1}{4}$

5. $\frac{2}{4}$ $\frac{5}{10}$

6. $\frac{3}{10}$ $\frac{1}{4}$

7. $\frac{1}{10}$ $\frac{1}{5}$ $\frac{1}{2}$

8. $\frac{1}{10}$ $\frac{1}{5}$ $\frac{1}{4}$

9. $\frac{1}{10}$ $\frac{1}{4}$ $\frac{1}{2}$

10. $\frac{2}{4}$ $\frac{3}{5}$ $\frac{9}{10}$

11. Answers will vary

Page 20

Comparing fractions

1. $\frac{1}{2} = \frac{2}{4}$

2. $\frac{1}{3} = \frac{2}{6}$

3. $\frac{2}{3} = \frac{4}{6}$

4. $\frac{6}{6} = \frac{3}{3}$

5. $\frac{1}{2} = \frac{3}{6}$

6. $\frac{3}{4} = \frac{9}{12}$

7. $\frac{3}{12} = \frac{1}{4}$

8. $\frac{4}{12} = \frac{1}{3}$

9. $\frac{8}{12} = \frac{2}{3}$

Rocket

1 whole									
$\frac{1}{2}$					$\frac{1}{2}$				
$\frac{1}{5}$		$\frac{1}{5}$		$\frac{1}{5}$		$\frac{1}{5}$		$\frac{1}{5}$	
$\frac{1}{10}$	$\frac{1}{10}$	$\frac{1}{10}$	$\frac{1}{10}$	$\frac{1}{10}$	$\frac{1}{10}$	$\frac{1}{10}$	$\frac{1}{10}$	$\frac{1}{10}$	$\frac{1}{10}$

Children's statements will vary.

10. $\frac{3}{6} = \frac{1}{2}$ $\frac{4}{8} = \frac{1}{2}$ Yes, they are giving away the same fraction.

Page 21

Comparing fractions

1. $\frac{1}{3} < \frac{1}{2}$

2. $\frac{2}{3} < \frac{3}{4}$

3. $\frac{3}{5} < \frac{5}{6}$

4. $\frac{1}{4} < \frac{2}{5}$

5. $\frac{3}{8} < \frac{1}{2}$

6. $\frac{4}{5} < \frac{7}{8}$

Rocket The first plate in question 6 has the most to eat. The second plate in question 4 has the least to eat. Answers as to how children know may vary.

7. $\frac{1}{2} < \frac{2}{3}$

8. $\frac{1}{3} > \frac{1}{4}$

9. $\frac{2}{3} < \frac{3}{4}$

10. $\frac{2}{4} = \frac{1}{2}$

11. $\frac{1}{4} < \frac{1}{2}$

12. $\frac{2}{3} > \frac{2}{4}$

13. $\frac{1}{3} < \frac{5}{12}$

14. $\frac{7}{12} < \frac{2}{3}$

15. $\frac{5}{6} > \frac{2}{3}$

16. $\frac{1}{6} < \frac{1}{3}$

17. $\frac{3}{6} = \frac{6}{12}$

18. $\frac{5}{6} < \frac{11}{12}$

Page 22

Ordering fractions

1. $\frac{1}{5}$, $\frac{1}{3}$, $\frac{2}{4}$, $\frac{3}{5}$, $\frac{2}{3}$

2. $\frac{1}{4}$, $\frac{1}{3}$, $\frac{2}{5}$, $\frac{2}{4}$, $\frac{4}{5}$

3. $\frac{2}{8}$, $\frac{1}{3}$, $\frac{3}{6}$, $\frac{3}{4}$, $\frac{7}{8}$

4. $\frac{1}{4}$, $\frac{2}{5}$, $\frac{1}{2}$, $\frac{3}{4}$, $\frac{4}{5}$

5. $\frac{1}{6}$, $\frac{3}{8}$, $\frac{2}{3}$, $\frac{3}{4}$, $\frac{5}{6}$

6. $\frac{1}{8}$, $\frac{3}{8}$, $\frac{2}{4}$, $\frac{5}{8}$, $\frac{2}{3}$

7. $\frac{2}{5}$, $\frac{1}{2}$, $\frac{3}{5}$, $\frac{2}{3}$, $\frac{3}{4}$

8. $\frac{1}{8}$, $\frac{1}{3}$, $\frac{2}{4}$, $\frac{5}{8}$, $\frac{2}{3}$

9. $\frac{1}{6}$, $\frac{1}{3}$, $\frac{3}{4}$, $\frac{4}{5}$, $\frac{7}{8}$

10. Answers will vary. Possible answers include:
$\frac{2}{3}$, $\frac{3}{5}$, $\frac{4}{6}$, $\frac{5}{7}$, $\frac{5}{8}$, $\frac{6}{9}$, $\frac{7}{12}$, $\frac{8}{12}$

Rocket Answers will vary.

Page 23

Ordering fractions

Fraction of red candles in each set:

1. $\frac{3}{4}$

2. $\frac{2}{3}$

3. $\frac{3}{5}$

4. $\frac{5}{6}$

5. $\frac{2}{5}$

6. $\frac{3}{7}$

7. $\frac{5}{8}$

8. $\frac{2}{7}$

9. $\frac{3}{10}$

Fraction of yellow candles in each set:

1. $\frac{1}{4}$

2. $\frac{1}{3}$

3. $\frac{2}{5}$

4. $\frac{1}{6}$

5. $\frac{3}{5}$

6. $\frac{4}{7}$

7. $\frac{3}{8}$

8. $\frac{5}{7}$

9. $\frac{7}{10}$

In order: $\frac{5}{7}$, $\frac{7}{10}$, $\frac{3}{5}$, $\frac{4}{7}$, $\frac{2}{5}$, $\frac{3}{8}$, $\frac{1}{3}$, $\frac{1}{4}$, $\frac{1}{6}$

Rocket $\frac{1}{2}$

$\frac{1}{3}$ $\frac{2}{3}$

$\frac{1}{4}$ $\frac{2}{4}$ $\frac{3}{4}$

$\frac{1}{5}$ $\frac{2}{5}$ $\frac{3}{5}$ $\frac{4}{5}$

$\frac{1}{6}$ $\frac{2}{6}$ $\frac{3}{6}$ $\frac{4}{6}$ $\frac{5}{6}$

The largest is $\frac{5}{6}$, the smallest is $\frac{1}{6}$.

Page 24

Comparing fractions

1. $\frac{1}{4}$

2. $\frac{3}{4}$

3. $\frac{3}{6}$

4. $\frac{4}{8}$

5. $\frac{6}{8}$

6. $\frac{1}{3}$

Rocket Answers will vary

7. 1, 2

8. 1, 2, 3, 4, 5

9. 1

10. 1, 2

11. 4, 5

12. 4

13. 5, 6, 7

14. 1, 2

15. 1

16. 1, 2

17. 4

18. 3

Fractions PPMs

PPM 137

Which are halves?

The children should circle the three shapes that show halves.

PPM 138

Finding halves

Children's answers may vary.

PPM 139

Shading halves

1. $\frac{1}{2}$ of 8 = 4
2. $\frac{1}{2}$ of 6 = 3
3. $\frac{1}{2}$ of 12 = 6
4. $\frac{1}{2}$ of 14 = 7
5. $\frac{1}{2}$ of 10 = 5
6. $\frac{1}{2}$ of 16 = 8

PPM 140

Doubling and halving

2. Half: 5   All: 10
3. Half: 6   All: 12
4. Half: 4   All: 8
5. Half: 8   All: 16
6. Half: 7   All: 14
7. Half: 9   All: 18
8. Half: 2   All: 4
9. Half: 10   All: 20

PPM 141

Halves

1. $\frac{1}{2}$ of 6 = 3
2. $\frac{1}{2}$ of 8 = 4
3. $\frac{1}{2}$ of 12 = 6
4. $\frac{1}{2}$ of 10 = 5
5. $\frac{1}{2}$ of 18 = 9
6. $\frac{1}{2}$ of 30 = 15
7. $\frac{1}{2}$ of 20 = 10
8. $\frac{1}{2}$ of 40 = 20
9. $\frac{1}{2}$ of 14 = 7
10. $\frac{1}{2}$ of 26 = 13
11. $\frac{1}{2}$ of 24 = 12
12. $\frac{1}{2}$ of 22 = 11
13. $\frac{1}{2}$ of 28 = 14
14. $\frac{1}{2}$ of 34 = 17
15. $\frac{1}{2}$ of 2 = 1
16. $\frac{1}{2}$ of 38 = 19

PPM 142

Which can you halve with none left over?

2. Yes
3. No
4. Yes
5. No
6. No
7. Yes
8. No

PPM 143

How many?

2. 5 bananas
3. $4\frac{1}{2}$ cherries
4. 9 cherries
5. $5\frac{1}{2}$ lemons
6. 11 lemons
7. $9\frac{1}{2}$ apples
8. 19 apples
9. $3\frac{1}{2}$ oranges
10. 7 oranges
11. $6\frac{1}{2}$ pineapples
12. 13 pineapples

PPM 144

What quantity?

1. 0, $\frac{1}{2}$, 1, $1\frac{1}{2}$, 2, $2\frac{1}{2}$, 3
2. 3, $3\frac{1}{2}$, 4, $4\frac{1}{2}$, 5, $5\frac{1}{2}$, 6
3. 6, $6\frac{1}{2}$, 7, $7\frac{1}{2}$, 8, $8\frac{1}{2}$, 9

PPM 145

Halves on a number line

1. $1\frac{1}{2}$, $2\frac{1}{2}$, 3
2. $\frac{1}{2}$, 2, $3\frac{1}{2}$, 4
3. $4\frac{1}{2}$, 5, $6\frac{1}{2}$
4. $8\frac{1}{2}$, 9, $9\frac{1}{2}$, $10\frac{1}{2}$, 11, $11\frac{1}{2}$

PPM 146

Discussion

Children's discussions.

PPM 147

Which are quarters?

These shapes show quarters:

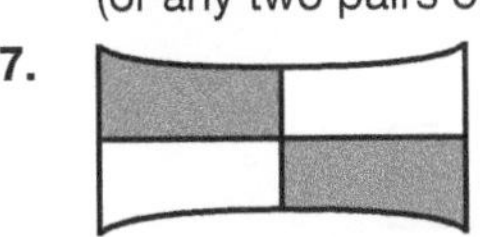

PPM 148

Quarters

1. 

(or opposite quarters)

2. 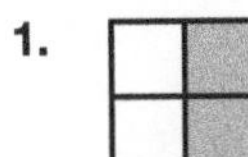

3. 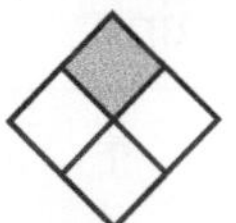

4. 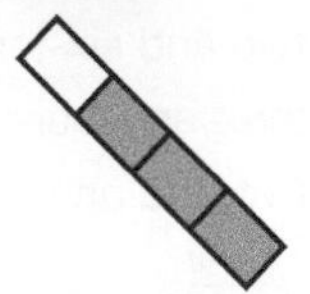

(or any two pairs of eighths)

5. 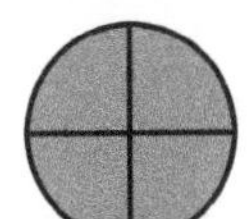

6. 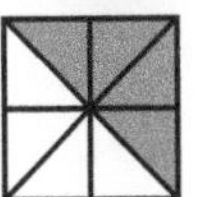

(or any two pairs of eighths)

7. 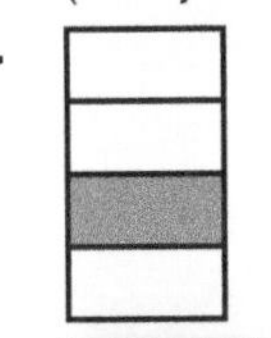

(or adjacent quarters)

8. 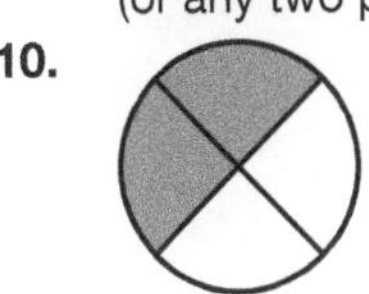

9. 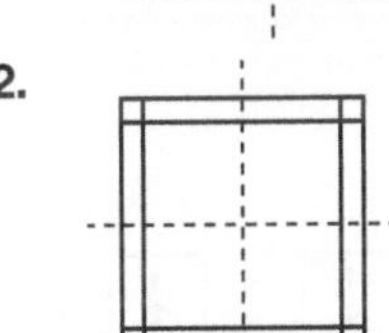

(or any two pairs of eighths)

10. 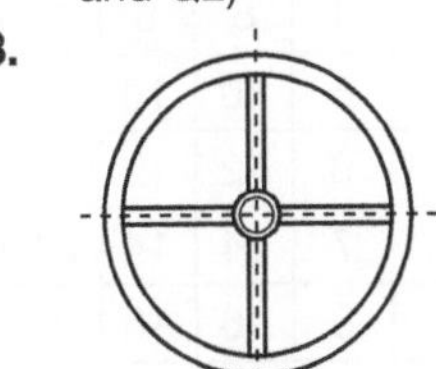

(or two opposite quarters)

PPM 149

Finding quarters

1.

2.

(diagonals can also be used in Q1 and Q2)

3.

4.

5.

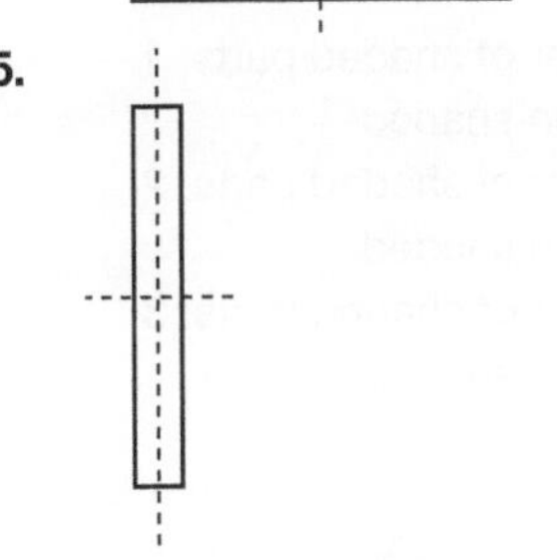

(or four equal lengths)

6.

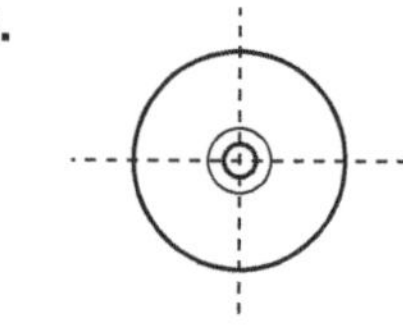

7.

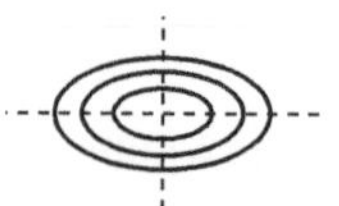

8.

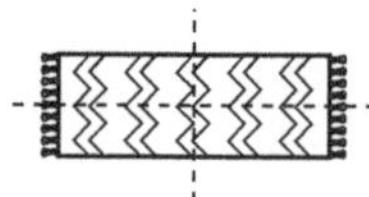

PPM 150
Fractions

1. $\frac{1}{4}$ of 8 = 2
2. $\frac{1}{4}$ of 12 = 3
3. $\frac{1}{4}$ of 20 = 5
4. $\frac{1}{4}$ of 24 = 6
5. $\frac{1}{4}$ of 4 = 1
6. $\frac{1}{4}$ of 16 = 4

PPM 151
Halves to quarters

Number of cubes	Half	Quarter
8	4	2
12	6	3
16	8	4
20	10	5
24	12	6
28	14	7
32	16	8

PPM 152
Quarters

Number of cubes	$\frac{1}{4}$	$\frac{2}{4}$	$\frac{1}{2}$	$\frac{3}{4}$
8	2	4	4	6
12	3	6	6	9
16	4	8	8	12
20	5	10	10	15
24	6	12	12	18
28	7	14	14	21
32	8	16	16	24

Children will notice that $\frac{2}{4} = \frac{1}{2}$.

PPM 153
Writing fractions

1. Number of shaded parts: 1
   Fraction shaded: $\frac{1}{4}$
2. Number of shaded parts: 2
   Fraction shaded: $\frac{2}{4}$
3. Number of shaded parts: 3
   Fraction shaded: $\frac{3}{4}$

PPM 154
How many?

2. $3\frac{3}{4}$ lemons
3. $4\frac{1}{4}$ apples
4. 1 orange
5. $5\frac{1}{2}$ lemons
6. $5\frac{3}{4}$ lemons
7. $\frac{3}{4}$ of an apple
8. $6\frac{1}{2}$ apples

PPM 155
What quantity?

1. 0, $\frac{1}{4}$, $\frac{2}{4}$ (or $\frac{1}{2}$), $\frac{3}{4}$, 1, $1\frac{1}{4}$, $1\frac{2}{4}$ (or $1\frac{1}{2}$)
2. 2, $2\frac{1}{4}$, $2\frac{2}{4}$ (or $2\frac{1}{2}$), $2\frac{3}{4}$, 3, $3\frac{1}{4}$, $3\frac{2}{4}$ (or $3\frac{1}{2}$), $3\frac{3}{4}$, 4
3. $7\frac{1}{2}$, $7\frac{3}{4}$, 8, $8\frac{1}{4}$, $8\frac{2}{4}$ (or $8\frac{1}{2}$), $8\frac{3}{4}$, 9

PPM 156
Quarters on a number line

1. $\frac{3}{4}$, $1\frac{1}{4}$, $1\frac{2}{4}$ (or $1\frac{1}{2}$)
2. 2, $2\frac{3}{4}$, $3\frac{1}{4}$, $3\frac{2}{4}$ (or $3\frac{1}{2}$), 4
3. $6\frac{1}{2}$, 7, $7\frac{2}{4}$ (or $7\frac{1}{2}$)
4. $9\frac{1}{4}$, $9\frac{2}{4}$ (or $9\frac{1}{2}$), $9\frac{3}{4}$, $10\frac{1}{4}$, $10\frac{2}{4}$ (or $10\frac{1}{2}$), $10\frac{3}{4}$

PPM 157
Tenths

1. $\frac{1}{10}$ of 60 = 6
2. $\frac{1}{10}$ of 80 = 8
3. $\frac{1}{10}$ of 10 = 1
4. $\frac{1}{10}$ of 50 = 5
5. $\frac{1}{10}$ of 100 = 10
6. $\frac{1}{10}$ of 100 = 10
7. $\frac{1}{10}$ of 200 = 20
8. $\frac{1}{10}$ of 70 = 7
9. $\frac{1}{10}$ of 120 = 12
10. $\frac{1}{10}$ of 120 = 12
11. $\frac{1}{10}$ of 160 = 16
12. $\frac{1}{10}$ of 180 = 18
13. $\frac{1}{10}$ of 400 = 40
14. $\frac{1}{10}$ of 200 = 20

PPM 158
Tenths on a number line

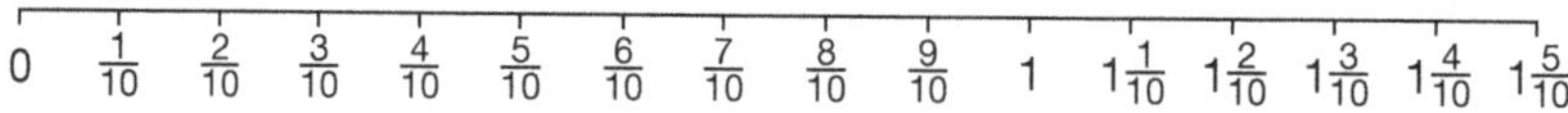

0, $\frac{1}{10}$, $\frac{2}{10}$, $\frac{3}{10}$, $\frac{4}{10}$, $\frac{5}{10}$, $\frac{6}{10}$, $\frac{7}{10}$, $\frac{8}{10}$, $\frac{9}{10}$, 1, $1\frac{1}{10}$, $1\frac{2}{10}$, $1\frac{3}{10}$, $1\frac{4}{10}$, $1\frac{5}{10}$

PPM 159
Wholes and tenths

2. $2\frac{6}{10}$ is shaded   two and six-tenths
3. $3\frac{4}{10}$ is shaded   three and four-tenths
4. $2\frac{1}{10}$ is shaded   two and one-tenth
5.

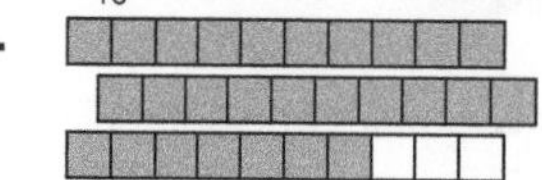

6.

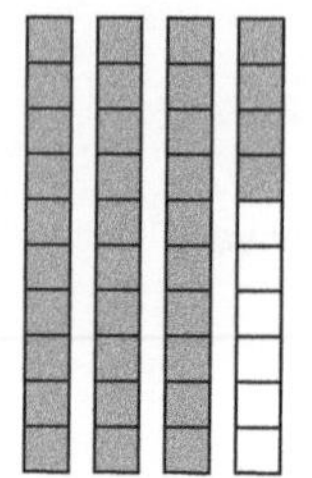

7.

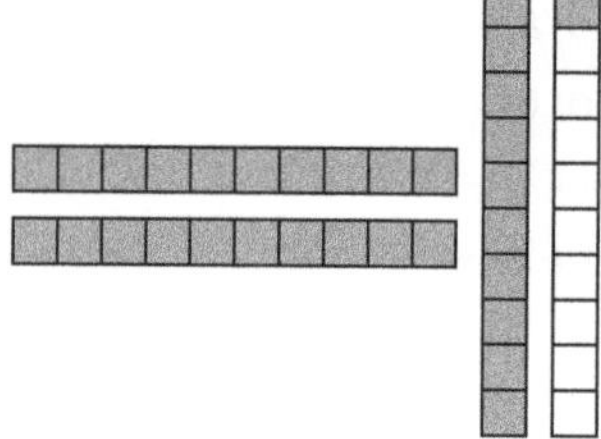

PPM 160
Fifths

2. $\frac{1}{5}$ of 15 = 3
3. $\frac{1}{5}$ of 25 = 5
4. $\frac{1}{5}$ of 40 = 8
5. $\frac{1}{5}$ of 30 = 6
6. $\frac{1}{5}$ of 45 = 9
7. $\frac{1}{5}$ of 25 = 5
8. $\frac{1}{5}$ of 50 = 10
9. $\frac{1}{5}$ of 35 = 7
10. $\frac{1}{5}$ of 20 = 4
11. $\frac{1}{5}$ of 10 = 2
12. $\frac{1}{5}$ of 5 = 1
13. $\frac{1}{5}$ of 0 = 0
14. $\frac{1}{5}$ of 1 = $\frac{1}{5}$

PPM 161
Fifths to tenths

1. 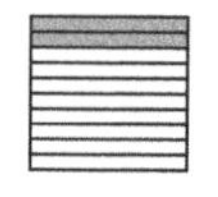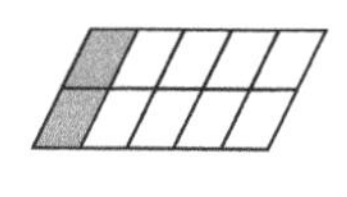
2. 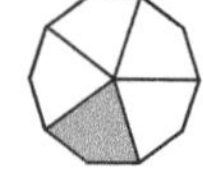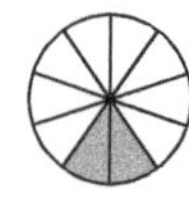
3. 
4. 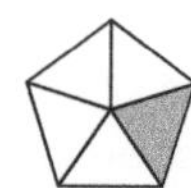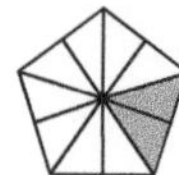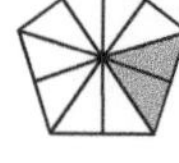

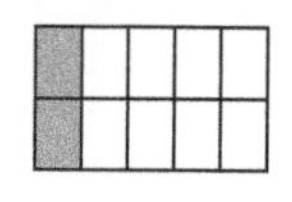

PPM 162

Fifths on a number line

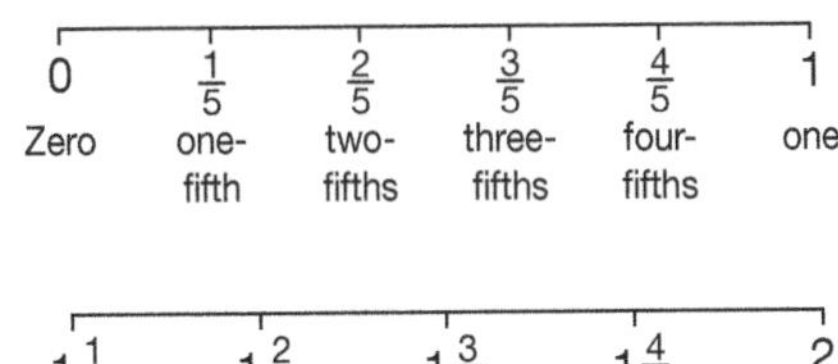

PPM 163

Wholes and fifths

2. $2\frac{3}{5}$ shaded   two and three-fifths
3. $1\frac{4}{5}$ shaded   one and four-fifths
4. 3 shaded   3 wholes
5. 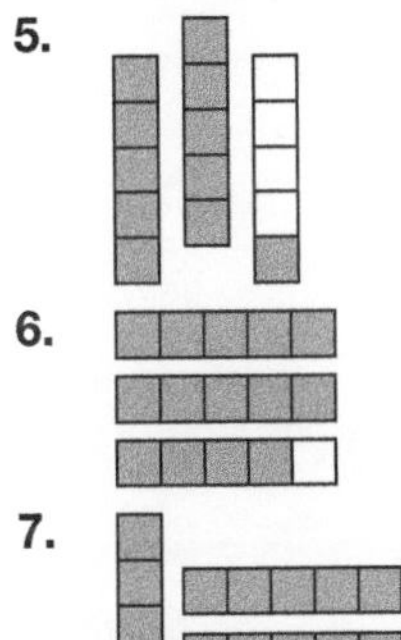
6.
7.

PPM 164

Fifths and tenths

1. $\frac{3}{10}, \frac{4}{10}, \frac{5}{10}, \frac{6}{10}, \frac{7}{10}, \frac{8}{10}, \frac{9}{10}, \frac{10}{10}$ (1)
2. $\frac{1}{5}, \frac{2}{5}, \frac{3}{5}, \frac{4}{5}, \frac{5}{5}$
3. Fifths must be entered on the top lines and tenths on the bottom lines

PPM 165

Fractions on a number line

1. $\frac{1}{2}$
2. $\frac{1}{3}, \frac{2}{3}$
3. $\frac{1}{2}, \frac{3}{4}$
4. $\frac{2}{5}, \frac{4}{5}$
5. $\frac{2}{10}, \frac{5}{10}, \frac{8}{10}$

PPM 166

Comparing many fractions

Every second section must be coloured with a different colour being used for each row.
False, false, false, true, true.

PPM 167

Comparing fractions

1. $\frac{1}{4} < \frac{1}{2}$
2. $\frac{1}{3} > \frac{1}{4}$
3. $\frac{2}{3} < \frac{3}{4}$
4. $\frac{1}{4} < \frac{3}{8}$
5. $\frac{5}{6} > \frac{3}{4}$
6. $\frac{5}{8} < \frac{2}{3}$
7. $\frac{7}{12} > \frac{1}{2}$
8. $\frac{8}{12} = \frac{2}{3}$
9. $\frac{5}{12} < \frac{3}{6}$
10. $\frac{3}{4} = \frac{9}{12}$
11. $\frac{2}{6} = \frac{4}{12}$
12. $\frac{3}{8} > \frac{1}{3}$

PPM 168

Fractions on a number line

1. $\frac{1}{4}, \frac{2}{5}, \frac{2}{4}, \frac{3}{5}, \frac{5}{5}$
2. $\frac{1}{4}, \frac{2}{5}, \frac{2}{4}, \frac{3}{4}, \frac{4}{5}$
3. $\frac{2}{8}, \frac{2}{6}, \frac{3}{6}, \frac{6}{8}, \frac{7}{8}$
4. $\frac{1}{4}, \frac{2}{5}, \frac{3}{4}, \frac{4}{5}, \frac{5}{5}$
5. $\frac{1}{6}, \frac{2}{8}, \frac{3}{8}, \frac{6}{8}, \frac{5}{6}$
6. $\frac{1}{8}, \frac{1}{4}, \frac{3}{8}, \frac{2}{4}, \frac{5}{8}$

PPM 169

Matching fractions

$\frac{1}{2}$ and $\frac{2}{4}$, $\frac{1}{3}$ and $\frac{2}{6}$, $\frac{1}{5}$ and $\frac{2}{10}$, $\frac{2}{3}$ and $\frac{4}{6}$, $\frac{2}{5}$ and $\frac{4}{10}$

$\frac{7}{10}$ and $\frac{2}{8}$ have no partner

Author team
Peter Gorrie, Lynda Keith, Lynne McClure and Amy Sinclair

Part of Pearson

Heinemann is an imprint of Pearson Education Limited, a company incorporated in England and Wales, having its registered office at Edinburgh Gate, Harlow, Essex, CM20 2JE. Registered company number: 872828

www.pearsonschools.co.uk

Heinemann is a registered trademark of Pearson Education Limited

Text © Pearson Education Limited 2010

First published 2010

18
10 9 8 7

British Library Cataloguing in Publication Data
A catalogue record for this book is available from the British Library

ISBN 978 0 435 03330 9

Typeset by Tech-Set Ltd, Gateshead
Cover design by Pearson Education Limited
Cover illustration by Volker Beisler © Pearson Education Limited
Printed and bound in Great Britain by Ashford Colour Press Ltd

Acknowledgements
Every effort has been made to contact copyright holders of material reproduced in this book. Any omissions will be rectified in subsequent printings if notice is given to the publishers.

Answer Book
First Level – Exploring Number

Heinemann Active Maths is the first activity-led maths programme created specifically for Scottish schools with the Curriculum for Excellence and Active Learning at its heart.

Active Learning, Active Teaching, Real Progression

This Answer book provides answers for:
- Pupil Book 1 (Number Processes)
- Pupil Book 2 (Addition and Subtraction)
- Pupil Book 3 (Multiplication and Division)
- Pupil Book 4 (Fractions)
- Practice Photocopiable Master activities
- Appropriate Activity Photocopiable Master activities.

FROM THE PUBLISHERS OF A HISTORY OF CONFLICT BRITAIN AT WAR MAGAZINE

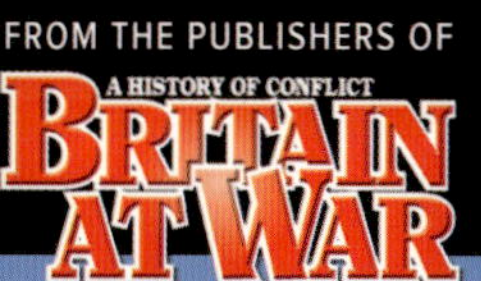

THE GUIDE TO THE SEABORNE ELITE

COMMANDO FORCE

BY SEA, BY LAND

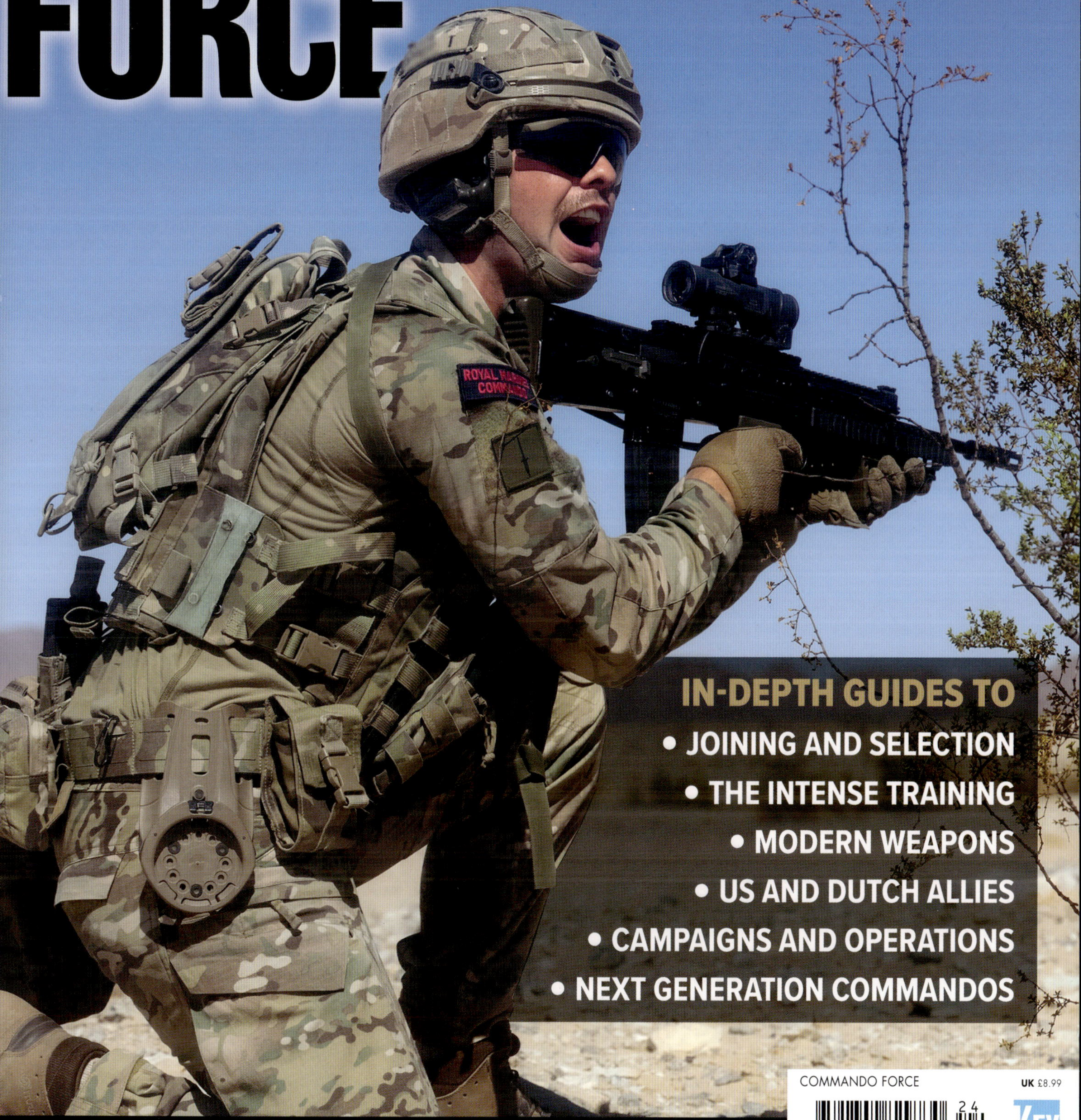

IN-DEPTH GUIDES TO

- JOINING AND SELECTION
- THE INTENSE TRAINING
- MODERN WEAPONS
- US AND DUTCH ALLIES
- CAMPAIGNS AND OPERATIONS
- NEXT GENERATION COMMANDOS

HIGH READINESS EXPEDITIONARY FORCES

COMMANDO FORCE UK £8.99

24

9 781802 829730

KEY Publishing

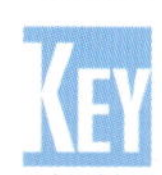

THE DESTINATION FOR
MILITARY ENTHUSIASTS
Visit us today and discover all our latest releases

FREE P&P* when you order from our online shop…
shop.keypublishing.com/specials
Call **+44 (0)1780 480404** *(Monday to Friday 9am - 5.30pm GMT)*
Free 2nd class P&P on all UK & BFPO orders. Overseas charges apply.